Variation Tolerant On-Chip Interconnects

ANALOG CIRCUITS AND SIGNAL PROCESSING

Series Editors:
Mohammed Ismail. The Ohio State University
Mohamad Sawan. École Polytechnique de Montréal

For further volumes:
http://www.springer.com/series/7381

Ethiopia Enideg Nigussie

Variation Tolerant On-Chip Interconnects

Ethiopia Enideg Nigussie
University of Turku
Turku
Finland
ethnig@utu.fi

ISBN 978-1-4614-0130-8 e-ISBN 978-1-4614-0131-5
DOI 10.1007/978-1-4614-0131-5
Springer New York Dordrecht Heidelberg London

Library of Congress Control Number: 2011943559

Printed on acid-free paper

Springer is part of Springer Science+Business Media (www.springer.com)

Preface

The design paradigm shift from single-core to multicore systems and from core-centric to interconnect-centric designs has emphasized the importance of high performance and reliable on-chip interconnects. In sub-100 nm technologies, variability has become a major challenge and it is considered one of the primary limiters for technology scaling. The inability to precisely control the manufacturing process leads to unpredictable device and wire characteristics, which in turn cause performance and power variability besides error-prone behavior. The performance and reliability of an interconnect is also affected by the environment in which it operates such as temperature, power supply voltage and noise. All these variations cause the signal propagation delay of the interconnect to be uncertain which in turn affects the performance and reliability of the communication significantly. Traditionally corner based analysis has been used to guard against yield loss resulting from these variations; however, with increasing number of sources of variation, corner based methods are becoming overly pessimistic and computationally expensive. Self-timed design methodologies can make the communication resilient to delay variations. More specifically, self-timed delay-insensitive links can operate correctly in the presence of delay variations in gates and interconnecting wires.

In this monograph designs of high performance and variation tolerant on-chip interconnects are presented. The design and implementation of these interconnects are based on formulation and integration of different circuit level techniques. Since delay variations are inevitable, the design focuses on self-timed delay-insensitive communication. In this regard, design and optimization of delay-insensitive data encoding/decoding schemes as well as formulation of efficient communication protocols are performed. To compensate the delay overhead of delay-insensitive communication, high speed signaling techniques are developed and implemented. In addition, a novel high speed completion detection technique is devised and implemented to solve the performance bottleneck caused by conventional completion detection methods. A high-throughput and power efficient serial interconnect is also designed in order to be used as a long-range on-chip communication link. Furthermore, an interconnect calibration technique after every power start-up of a system is developed and implemented to ensure signal integrity of the interconnects despite process, wearout and aging caused variations. A runtime supply voltage and temperature (VT) variation tolerance technique is also devised and implemented for the interconnects. These Process, Voltage and Temperature (PVT) variation tolerance schemes make the interconnects adaptive to the effect of variations, enabling continuous and reliable operation of the interconnect.

The manuscript is organized as follows. The introduction in Chap. 1 focuses on the drive for interconnect-centric design and challenges of global on-chip communication. In Chap. 2, the design techniques used to implement the presented high performance delay-insensitive interconnects are discussed. Methods and basis for estimating wire parasitics along with the electrical level modeling of wires is discussed in Chap. 3. Design and analysis of the three delay-insensitive current sensing on-chip interconnects are presented in Chap. 4. In addition, analysis of their performance and power consumption as well as comparison with conventional delay-insensitive on-chip interconnects are presented. In Chap. 6, a high speed completion detection technique as well as its design is presented in order to enhance the performance of the delay-insensitive interconnects. Furthermore, two of the interconnects presented in Chap. 4 are redesigned and presented as case studies to demonstrate the advantage of the presented completion detection technique. Analysis of their performance, energy dissipation and area besides comparison with the reference cases are also discussed. In Chap. 6, implementation and analysis of high-throughput serial on-chip interconnect targeted for long-range communication is presented. Also, comparison of throughput, energy and area between fully bit-parallel, bit-serial and semi-serial links are performed. All the interconnects which are presented in Chaps. 4–6 are redesigned using 65 nm CMOS technology and their performance, energy dissipation, and area are compared in Chap. 7. In Chap. 8, circuit techniques as well as implementations to tolerate process, supply VT variation effects on the signal integrity of the interconnects are presented.

Although much care has been made in the preparation of the manuscript, flaws and errors might still exist due to erring human nature. Suggestions and appropriate comments are highly appreciated.

Turku, Finland Ethiopia Enideg Nigussie

Acknowledgements

I would like to take this opportunity to express my sincere gratitude to the people and institutions that have helped me to accomplish this research work. This manuscript is developed from my doctoral dissertation and due to this I am grateful to my doctoral research supervisors Adj. Prof. Juha Plosila, Prof. Jouni Isoaho, and Prof. Hannu Tenhunen. Their inspiration, guidance and support has been the main driving force for this research. The support from the Department of Information Technology, University of Turku, Finland where I carried out this research is gratefully acknowledged.

My heartfelt appreciation goes to Professor Mohammed Ismail, the series editor of Analog Circuits and Signal Processing, for his comments on the contents of the manuscript as well as for inviting me to write this monograph.

Many thanks to the editorial staff of Springer, especially Charles B. Glaser, Senior Editor–Electrical Engineering, they have been wonderfully supportive and encouraging.

A large dept of gratitude is owed to my wonderful mother Yisgedu Agonafir. Though I lost you many years ago, you are still inspiring me to work hard and reach further.

Contents

Chapter 1
Introduction

The continuous development of semiconductor technology over the last five decades has been the enabling factor that has driven many huge changes in our everyday life. Personal computing, mobile communications, Internet, broadband technology and automobile industry, are obvious examples. This remarkable development is the result of technology scaling that led to fabrication of Integrated Circuit (IC) with smaller feature sizes, higher levels of integration and faster operating frequencies. The process of device scaling evolved from few micrometers to nanometers today, and the circuit complexity has advanced from Small-Scale Integration (SSI) in 1960s to Giga-Scale Integration (GSI) in 2000s. It is predicted that this integration continues at a faster speed towards a trillion transistors per chip, Tera-Scale Integration (TSI) era, in 2020s. Today, not only digital devices and memories, but also analog/mixed-signal blocks, MEMS based sensors, and other functional blocks are being integrated on the same die to build a complete system. However, the benefits of system integration are significantly reduced without efficient communication between these blocks. Thus, this book addresses the problems of global on-chip interconnects using novel circuit level techniques.

1.1 Emergence of Interconnect-Centric Design

The performance of transistors is continually improved through scaling. However, the impact of technology scaling on long wires is reverse. In order to tackle this problem, there is a shift in system design approach from computation-centric to interconnect-centric. Interconnect is used to distribute clock and signals and to provide power and ground to and among many functional blocks on a chip. The increase in die size due to increasing chip functionality makes it more difficult to deliver signals across the chip in one clock cycle [14]. These emphasizes the importance of interconnect-centric design to optimize the overall chip performance in nanometer technologies.

E.E. Nigussie, *Variation Tolerant On-Chip Interconnects*, Analog Circuits and Signal Processing, DOI 10.1007/978-1-4614-0131-5_1,

1.1.1 Device and Interconnect Scaling

The primary goals of technology scaling are decreasing gate delay, increasing gate density, and reducing energy per storing/operating [1]. At the moment, the feature size is scaling down at a rate of 0.7 per year [110] in compliance with Moore's Law [2]. This decreases the gate delay by 30%, doubles the gate density and reduces the energy per switching by 65%. As the history of integration density reveals, in 1960 one transistor consisted of 10^{20} atoms in a volume of 0.1cm^3 and in 2000 these number were 10^7 atoms in $0.01\,\mu\text{m}^3$, leading to a higher capacity of integration. Similarly, the energy for storing/operating 1 bit is reduced because the energy required for charging and discharging capacitors is lowered due to the reduction in capacitors area from 1cm^2 to $0.01\,\mu\text{m}^2$ and furthermore the supply voltages are scaled down from 10 V to 1 V [3].

Scaling down of transistor's dimensions leads to improvements in both transistor cost and performance. However, scaling down of interconnect cross-sectional dimensions degrades performance. The ideal scaling of interconnect assumes that the width and height of the wire are reduced with the same scaling factor as gates' dimensions, leading to taller and narrower wires. As a result, the resistance of a unit length wire increases at the rate of 104% per year. The length of local wires scales the same way as the logic, whereas global wires tend to track the chip dimensions. In general, die area should decrease by 50% per year in successive technology but new designs integrate more transistors and functionality per chip, resulting in a need for die area increment. Die area has been increasing 13% per year. Consequently global interconnect length increases at a rate of 6% per year, and its RC time constant increases by approximately 130% per year.

To mitigate the increase in wire delay, various techniques have been developed from both geometric structure and materials perspective, such as high aspect ratio, multiple-layer metallization, copper technology and low-k dielectrics. A higher aspect ratio along with smaller wire pitch leads to a reduction in RC delay but this approach has two problems. First, manufacturing of lines and vias with aspect ratio larger than 4 becomes unreliable due to the difficulty in filling a deep and narrow trench completely with metal. Second, the increase in line thickness due to a higher aspect ratio results in a larger coupling capacitance to neighboring wires, which increases both the RC delay component and the signal coupling noise. These two undesirable effects limit the practicality of this technique for nanometer technologies. In multiple-layer metallization approach, different scaling factors for local and global interconnects are used to satisfy the need for higher density, reduced RC delay, and smaller resistive loss. In order to match the device density on the substrate and maintain the RC delay, the pitches and length of local wires scales at a much faster rate than the vertical dimensions. For global wires, the scaling is determined by the length of the chip edge and as a result signal delay on global wires increases continuously from one technology generation to the next. This reverse scaling of global interconnect is an undesirable consequence of technology scaling. From materials perspective two major advances have been

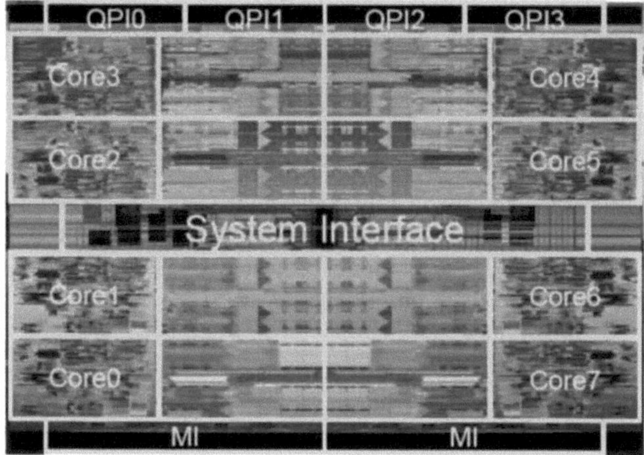

Fig. 1.1 Intel's 45 nm 8-core Xeon-EX processor [5]

done, the change in metal material from aluminum to copper and the introduction of low-k dielectrics to replace silicon oxide (SiO_2). However, in nanometer technology, all these techniques are insufficient to achieve the needed high-speed global on-chip communication. Hence, requiring additional techniques and other approaches.

1.1.2 System-on-Chip and Multicore Systems

The continued scaling of the semiconductor technology creates the potential of System-on-Chip (SoC) integration, that is, the integration of a complete electronic system including all its periphery and its interfaces to the outside world on a single die. SoC consists of several heterogeneous components with different implementation styles such as programmable processors, dedicated hardware to perform specific tasks, on-chip memories, input-output interfaces, and on-chip communication architecture that serves as the interconnection fabric for communication between these components. For instance, Intel's 45 nm Xeon® EX processor (Nehalem-EX) is a SoC which has eight 64-bit cores and a 24 MB shared L3 cache (Fig. 1.1). At the top stripe it has four Quick Path Interconnect (QPI) links, while the bottom stripe houses the Scalable Memory Interconnect (SMI) links. It also has a system interface that includes two memory controllers, two hub interfaces to the last level cache, an 8-port router, the power control unit (PCU) and the DFX control box. Usually such building blocks can be shared and also re-used as Intellectual Property (IP) blocks, which further improves the productivity and reduces time-to-market.

Today, SoC combines a diverse set of components using adaptive circuits, integrated sensors, sophisticated power-management techniques and increased parallelism to build products that are multicore and multi-function. Examples of such prototypes are Intel's 80 core TeraFLOPS [69], 167-processor computational platform [64] and FAUST chip (a reconfigurable baseband platform consisting of 23 computing units that can be configured to support the functions of specific baseband processing) [88]. This trend will continue and it will open up the feasibility of a wide range of applications, such as data mining, visual computing, and recognition, making use of massive parallel processing and tightly interdependent processes which brings into front the underlying interconnection capability. The interconnection between SoC components should provide reliable routing of data from the source to destination. It must also be able to guarantee latency or bandwidth to ensure that the application performance constraints are met.

1.1.3 Network-on-Chip

The increasing number of IP cores that can be integrated on a single chip enables implementation of complex applications using the SoC approach. The huge communication demands of these applications and the abundant computation power available on-chip put tremendous pressure on the communication architecture. Consequently, scalable communication architectures are needed for efficient implementation of SoC. Simple on-chip communication solutions do not scale up when the number of processing and storage arrays on a chip increases. For example, on-chip buses can serve a limited number of units, and beyond that performance degrades due to the bus parasitic capacitance and the complexity of arbitration. Network-on-Chip (NoC) is a communication infrastructure targeted for SoC consisting of tens or hundreds of resources. NoCs are an attempt to scale down the concepts of large-scale networks, and apply them to SoC domain. It separates the concerns of communication from computation by building on-chip communication structure. Each component of a SoC is viewed as a node of the on-chip communication network. NoC use packets to route data from the source to the destination component, via a network fabric that consists of switches(routers) and point-to-point links, which connect the resources to routers as well as the routers to each other to form a network. NoC provides better scalability than on-chip buses because as more resources are introduced to a system, also more routers and links are introduced to connect them to the network. The additional links and routers provide the communication capacity needed for the new resources. Many NoC realizations inherently contain some redundancy in the communication media, which can be used to provide a higher reliability and traffic balancing. This is in contrast to bus structures, which rely on a single communication medium.

NoC has been initially proposed as a design paradigm for on-chip communication in the beginning of this millennium [15–17, 44, 79]. Today, there are NoCs in commercial use such as *Arteris*™ [139] and *STNoC*™ [140, 141] as well as industrial products which use NoC as a communication backbone. The *TILE64*™ 64-core processor from Tilera [18, 66] and the 80-core Intel's TeraFLOPs processor [69] are recent examples of industrial products which proves the feasibility and potential of NoC. The interconnects designed and presented in this thesis can be used as a link between two NoC routers.

1.2 Challenges of Global On-Chip Interconnect

On-chip interconnect has become a primary challenge for high performance high complexity SoCs. Transmitting clock, data, and communication signals over large die areas requires long interconnections among the various circuit modules. As technology scales, the interconnect cross section decreases while operating frequencies increase. The impact of these trends on high performance systems is significant. Long interconnects with smaller cross sections exhibit increased capacitance and resistance, resulting in larger power consumption, and higher latency. Furthermore, wire inductance can no longer be ignored due to high signal frequencies and long wire lengths. The increasing number of cores per chip also places a premium on high-bandwidth, low-latency and low-power links between cores.

1.2.1 Performance and Power Consumption

The higher wire resistance, the increase in wire length and reduced wire spacing cause the global wire delay to increase considerably compared to the gate delay [110]. The gap between global interconnect delay and gate delay increases with technology scaling as can be seen from Fig. 1.2. Furthermore, the increase in die size due to increasing chip functionality makes it more difficult to deliver signals across the chip in one clock cycle [14]. According to the prediction of International Technology Roadmap for Semiconductors (ITRS), at 45 nm technology node, the RC delay is 542 ps for a 1 mm long minimum pitch copper global wire, whereas the clock frequency will reach 10 GHz (equivalent to 100 ps cycle time). The conventional approach to deal with this is to use pipelining in global signals, which increases the latency and power consumption when routing signals across functional blocks. Also, the total on-chip wire length will increase linearly with technology, reaching about 2.22 km cm^{-2} by the year 2010 [110]. This trend supports the assumption that long interconnects will be significant in future technologies. To combat these phenomena, traditional repeater insertion methods have been widely developed and adopted. Unfortunately, as interconnect lengths increase, the required number of repeaters increases tremendously. This results in significant power dissipation, increased delay, and larger area.

Interconnects, especially the global wires have also become a major source of power consumption. It has been reported that wire capacitance can take up to 70% of the total chip capacitance in contemporary designs [6]. Moreover, a rapid increase in chip operating frequencies further exacerbates the amount of dynamic

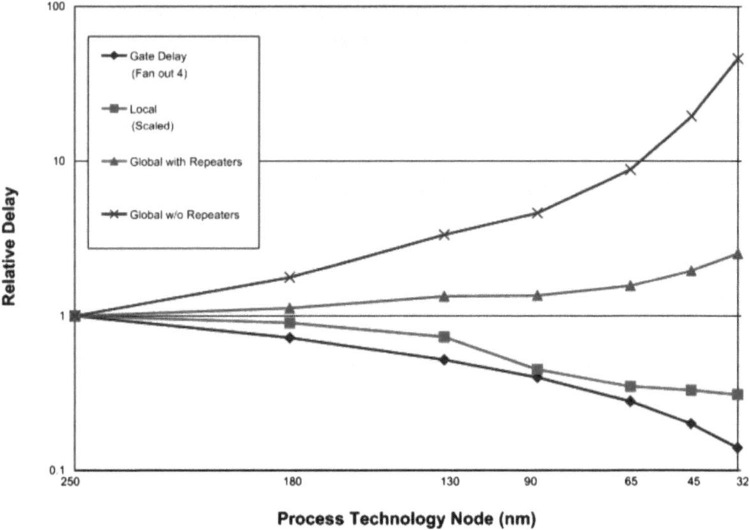

Fig. 1.2 Delay comparison [110]

power dissipated in the interconnect. Magen et al. found that interconnection power accounts for half the total dynamic power of a 130 nm microprocessor, and nearly 50% of the interconnect power is consumed by global wires [7]. With a projection that without changes in design philosophy, in the next five years up to 80% of microprocessor power will be consumed by interconnect.

As discussed in Sect. 1.1.3, NoCs have emerged as the seemingly best candidate to connect the cores on present and future SoCs. Latency and power consumption of NoC are among the critical challenges and need to be addressed at all abstraction levels [161]. The latency of networks is too large, leading to performance degradation when they are used in high performance systems. The power consumption of NoC implemented with current techniques is too high, by a factor of 10, to meet the expected needs of future SoCs. In NoC the network interconnects consume a significant part of the total power budget. For example, in TeraFLOPS [69] the network consumes up to 39% of the total chip power (76 W when operating at 5.1 GHz) [8]. 17% of the network power is consumed in the links (13 W at 5.1 GHz). Therefore, more emphasis should be put on circuit techniques that increase signal velocity on channels and reduce the power consumption of the interconnects.

1.2.2 Variability and Reliability

Variability has become a major challenge for designs in sub-100 nm technology nodes and it is considered one of the primary limiters for technology scaling [9–12, 118]. It is affecting device as well as interconnecting wire parameters. The inability to precisely control the manufacturing process leads to unpredictable device and wire characteristics, which in turn cause performance and power variability besides error-prone behavior [13, 118–130]. According to ITRS, within a few years delay and power variability reaches 63% and 76%, respectively [115]. In addition, systems performance is also affected by the environment in which it operates such as temperature, power supply voltage and noise.

The performance of an on-chip interconnect is determined by the electrical characteristics of the signaling circuit's devices and the interconnecting wire parasitics. From a process perspective, almost all manufacturing phases, etches, thin-film deposition, hot processes, and even wafer clean processes, influence device parameters and thus contribute to variabilities. Increased process complexity related to subwavelength lithography, chemical-mechanical polishing (CMP), and the implementation of low-k dielectrics leads to higher variability of wire resistance and capacitance. Variations in operating environment, spatial as well as temporal effects, can also have a similar impact. For example, the effective supply voltage of a transistor may vary across the chip due to changes in the voltage drop along the power grid. The local operating temperature of a transistor is affected by local variations in power dissipation. Crosstalk, resulting from capacitive and inductive coupling, could severely affect the timing and signal integrity of an interconnect. Because each victim wire experiences a different capacitive coupling length or a different inductive-coupling return path, the interconnect exhibits varying signal propagation delays under different switching patterns. All these variations cause the signal propagation delay of the interconnect to be uncertain which in turn affects the performance and reliability of the communication significantly.

Traditionally corner based analysis has been used to guard against yield loss resulting from these variations; however, with increasing number of sources of variation, corner based methods are becoming overly pessimistic and computationally expensive. Self-timed design methodologies can make the communication resilient to delay variations. More specifically, self-timed delay-insensitive links can operate correctly in the presence of delay variations in gates and interconnecting wires.

1.3 Global On-Chip Communication Techniques

SoC consists of many IP blocks. The different functions among different SoC blocks naturally cause them to work in different clock rates for optimal performance. Hence, coordination and communication between these components become challenging. Globally Asynchronous Locally Synchronous (GALS) scheme has been proposed as a solution. The idea of GALS is to partition a system into separate clock domains, which run at different clock rates, and the separated domains communicate with each other in an asynchronous manner.

1.3.1 GALS Communication

Globally synchronous communication is a thing of past because it is difficult to design with growing chip sizes, clock rates, relative wire delays and parameter variations. Moreover, high speed global clocks consume a significant portion of system power budgets and lack the flexibility to independently control the clock frequencies of submodules to achieve high energy efficiency. GALS facilitate the integration of independently designed blocks operating at different frequencies as well as fast block reuse by providing wrapper circuits to handle the inter-block communication.

NoCs with GALS clocking styles have been used in many proposed network designs and are expected to be an attractive approach to overcome many of the timing problems [79]. GALS simplifies clock tree design and results in easily scalable clocking systems. It also allows better energy savings since each functional unit can easily have its own independent clock and voltage [80]. Furthermore, it enables easy implementation of distributed power management system for the entire chip [81].

Generally there are two different implementation of GALS NoC: fully asynchronous (self-timed) and multi-synchronous. In self-timed GALS NoC, IP blocks use locally generated clock and there is a synchronous $\Leftrightarrow$ asynchronous interface between the network and the synchronous IP. Clockless networks such as MANGO [86], ANoC [87], ALPIN [81], FAUST chip [88], and QNoC [89] are examples of self-timed NoC. A systematic comparison between these two implementations shows that the self-timed network gives better saturation threshold, smaller average power consumption, slightly higher maximal bandwidth and 2.5 times smaller packet latency than the multi-synchronous implementation [58]. Furthermore, the risk of metastability introduced by the multiple bi-synchronous FIFOs used in the multi-synchronous implementation can be a critical issue. This risk is much lower in the self-timed approach since the metastability is entirely confined in the synchronous $\Leftrightarrow$ asynchronous interface. Due to these advantages, the focus of this book is on the design of self-timed delay-insensitive interconnects.

1.3.2 Self-timed Delay-Insensitive Communication

Delay-insensitive codes have been used in many applications for error detection and delay-insensitive communication. Their main feature is the ability of allowing the correct interpretation of the code word independently of the delay of individual bits. Hence, self-timed delay-insensitive data transfer is one of the most promising approaches to deal with delay uncertainties in on-chip interconnects. A self-timed delay-insensitive communication link assumes nothing about the delays in the wires and devices except that they are finite and positive, and therefore the reliability of communication is unaffected by the delay variations. Several delay-insensitive coding schemes have been proposed, but effective Complementary MetalOxideSemiconductor (CMOS) implementations are needed in order to make feasible self-timed on-chip interconnect. Dual-rail and 1-of-4 codes are well known and mostly used for on-chip delay-insensitive communication [59].

The delay insensitivity feature does not come free of cost, it has delay and area overhead. The data encoding at the transmitting side and decoding at the receiver side, as well as completion detection cause additional delay to the communication. In this thesis, different performance enhancement techniques for delay-insensitive interconnects are developed and implemented in order to compensate the delay overhead and achieve high performance communication.

1.4 Related Work

Increasing attention is placed on the design of on-chip interconnects due to the dominant limitation of global interconnect signal delays, power dissipation and delay uncertainty on overall system performance and reliability. It is imperative that future on-chip interconnect designs overcome these challenges. The conventional technique to improve the interconnect delay bottleneck is to insert repeaters by breaking the wire into several sections [111, 154]. Usually these wire sections are highly capacitive and high strength repeaters are needed. The adverse effect of this is increased power consumption; it has been estimated that over 50% of the power in a high performance microprocessor is dissipated by repeaters charging and discharging interconnects [6, 7]. The other approach is inserting register pipelines [156–159] to increase the communication throughput. This approach increases the latency of the communication, and furthermore, the number of registers needed increases with the size and the complexity of the system. This in turn increases the power consumption. These show that the conventional solutions are inadequate to meet the overall performance requirements of high performance electronic systems in nanometer regime.

1.4.1 High Performance Interconnect

In order to solve the delay and power problems of global interconnects, several alternatives have been proposed by ITRS [63]. Using different signaling methods is among the proposed alternatives. This approach utilizes available technology with innovative approaches to signaling and circuit operation to implement high speed global interconnects. In [40], high speed on-chip signaling method that relies on differential current-mode sensing to improve both delay and energy dissipation has been proposed and implemented. High speed and power efficient on-chip interconnect has been demonstrated using current mode signaling along with circuit techniques in [33]. In [39], a 10 Gbps/channel on-chip signaling system has been fabricated in 90 nm technology. It consists of current mode logic driver and receiver, and differential transmission line. By using impedance-unmatched driver it saves the energy per bit by 21% compared with a conventional impedance-matched driver. Energy-aware differential current sensing signaling through the use of differential leakage-aware amplifier has been proposed in [142]. A method to propagate signal near the speed of light has been demonstrated in [34], though the interconnect has high power consumption. Wave-pipelining has also been proposed for on-chip interconnects as a means to increase throughput [82, 108, 143, 155]. There are researches which focus on using signal conditioning, and high-speed transceivers in order to improve interconnect throughput [144–148]. All of these works concentrate in achieving high performance communication without dealing properly the reliability problem due to delay variations.

1.4.2 Variation Tolerant Interconnect

In nanometer scale technologies sources of variability are increasing and unavoidable, which creates several challenges in building reliable systems. Variability causes signal propagation delay uncertainty in interconnects which in turn causes error. Due to this, variation tolerant on-chip interconnects are needed. Self-timed design methodologies can make the interconnect robust to delay variations. Different self-timed delay-insensitive interconnects have been proposed in [37, 60, 61, 112, 113, 150–153, 160]. Most of these works concentrate only in the delay insensitivity feature and ignore the need for high performance interconnects. For example, the work in [37, 60, 61, 112, 150, 153, 160] use four-phase handshaking, which requires four traversals of the long wire per each data transfer. This decreases the performance of such interconnects significantly. In [150], delay-insensitive encoding, which minimizes the wiring overhead has been proposed but the encoder, decoder and completion detection logic complexity increases which increases the delay overhead and consequently reduces the throughput of the communication. An asynchronous DI interconnect which uses two-phase dual-rail encoding has been implemented and compared with synchronous interconnect in [113]. To improve the throughput of this interconnect locally clocked pipeline stages have been inserted, which increase both power consumption and the required area.

1.4.3 High Performance and Variation Tolerant Interconnect

The aim of this book is to formulate design techniques which enable both high performance and variation tolerant on-chip communication. The designed interconnects use two-phase handshaking and self-timed delay insensitive data transfer. To compensate the delay overhead due to delay-insensitive encoding, decoding and completion detection, different high-speed signaling techniques have been implemented. In addition, to minimize the delay overhead due to completion detection of wide bit transmissions, bit width insensitive high-speed completion detection technique has been developed and implemented. Furthermore, self-calibration, monitoring and reconfiguration techniques have been developed to guarantee the signal integrity of the interconnects despite PVT variations.

Chapter 2
Interconnect Design Techniques

In this chapter, power efficient design techniques for the delay-insensitive global on-chip interconnects are presented. It is a foundation work for the rest of the chapters. The chapter starts with the handshaking protocols and is followed by discussion of data encoding, decoding, and completion detection techniques. Furthermore, customized and advanced signaling techniques are also explained in detail.

2.1 Handshaking Protocols

In self-timed on-chip communication, handshaking protocol is used to transmit data between a sender and a receiver. The sender delivers data onto the channel and the receiver accepts data from the channel. The communication parties can be further classified as follows: active part initiates the data transfer and passive part responds to the active. The transfer direction is determined by the communication protocol.

The exchange of data through a channel is negotiated between the sender and the receiver using a handshaking protocol. For every data transfer, a request is transmitted by the active module which indicates the validity of the data on the channel. An acknowledgment transmission from the passive module indicates data acceptance and readiness of the receiver to accept the next data. The transmission of request and acknowledgment signals may occur on dedicated signaling wires or may be implicit in the data depending on the data encoding technique used but in either case, one event indicates data validity and the other data acceptance. The flow of data relative to the request event determines whether the channel is a push or pull channel. In a push channel data flows in the same direction as the request whereas in the pull channel data flows in the same direction as acknowledgment signal. These two types of channels are illustrated in Figs. 2.1 and 2.2. The push channel is assumed in all of the interconnects designed in this book.

In general, the request and acknowledgment signals may be transmitted using one of the two protocols described below; a two-phase transition-based handshaking also called a non return-to-zero protocol or a four-phase level-based protocol

E.E. Nigussie, *Variation Tolerant On-Chip Interconnects*, Analog Circuits and Signal Processing, DOI 10.1007/978-1-4614-0131-5_2,

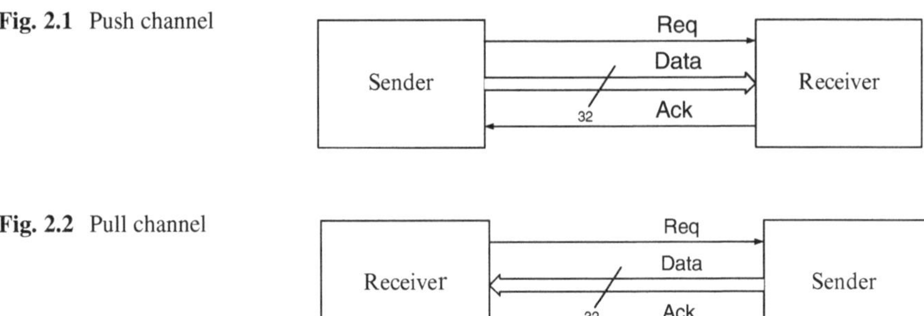

Fig. 2.1 Push channel

Fig. 2.2 Pull channel

(a return-to-zero scheme). There are other customized protocols such as single-track [19, 20] and one-phase [21–23]. The advantage of single-track handshaking is that it requires just two transitions per data transfer as opposed to four in four-phase protocol and avoids the requirement for event-triggered logic circuits of the two-phase protocol. However, the implemented circuit will run correctly only if it is not exposed to heavy ambient noise because single-track protocol relies momentarily on high impedance states on wires. The one-phase protocol requires only one communication action between the sender and the receiver which makes it faster than both two and four-phase handshaking. It uses a data coloring scheme to indicate data validity and acceptance. The transmitted symbol consists of both bit value and color information. There is a color detector circuit at both the transmitter and the receiver. The detector detects the signal in the wire and extracts the color information. The receiver accepts the data and changes its color when the color in the wire is the same as its own (data is valid). Also the transmitter sends the next data with new color after its color is the same as the wire color (data acceptance). Due to color detection at both sides, there is no need to transmit either the data validity or acceptance to one another. Even if it is attractive in saving communication time, it requires complex circuits and incurs additional power consumption.

The four-phase handshaking protocol, shown in Fig. 2.3, uses signal levels to indicate the validity of data and its acceptance by the receiver. That is the sender issues data and sets the request high, the receiver absorbs the data and sets the acknowledgment high. The sender responds by setting the request low (at this point data is no longer valid) and the sender acknowledges this by setting the acknowledgment low. This requires two transitions per data transfer on both request and acknowledgment signals. With increasing wire delay due to the reverse scaling effect of long wires, accommodating the return-to-zero phase leads to a significant reduction in throughput which makes the four-phase handshaking unattractive for global on-chip communication.

Unlike the four-phase protocol the signal levels are unimportant in the two-phase handshaking protocol where the information is carried by the transition. Both rising and falling transitions are equivalent, each being interpreted as a handshaking event. A push channel that uses the two-phase protocol passes data using a request signal transition, and acknowledges data reception with an acknowledge signal transition.

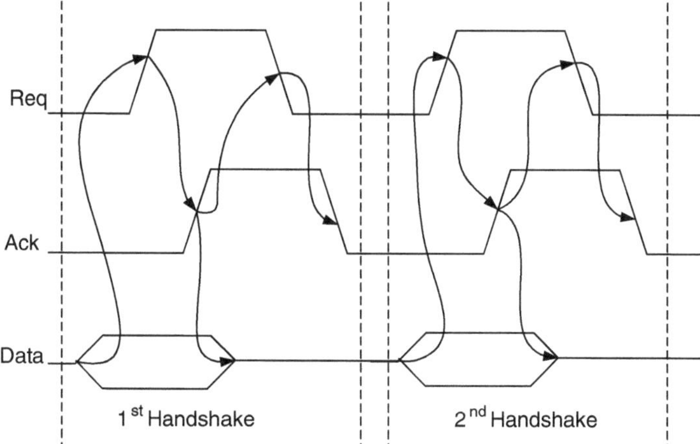

Fig. 2.3 Four-phase handshaking protocol

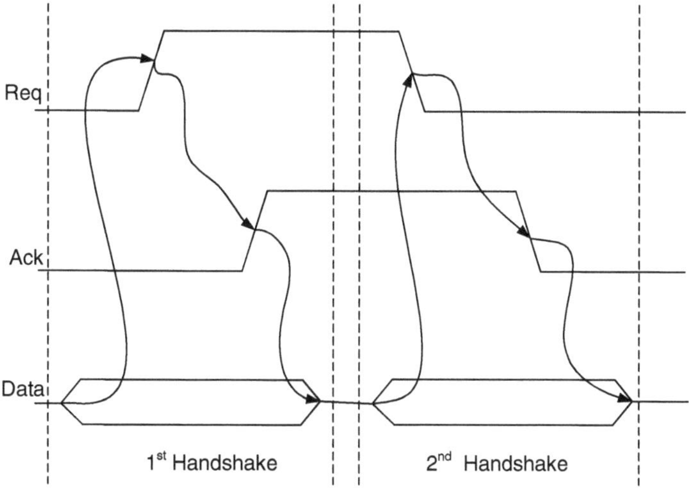

Fig. 2.4 Two-phase handshaking protocol

Two-phase handshaking is preferred for long on-chip communication since it reduces the required number of transitions by half and avoids the requirement of spacer compared to four-phase signaling [25]. This saves communication time and energy of the system significantly. Figure 2.4 illustrates the two-phase protocol.

Some may argue that the two-phase communication requires edge sensitive control logic circuits, which leads to considerable delay overhead. But the delays of edge sensitive logic circuits are much smaller than the global wire delay in nanometer CMOS technologies, which makes the use of two-phase handshaking advantageous over the four-phase. For example, in 65 nm CMOS technology, global wire delay is nine times the gate delay [110].

2.2 Data Encoding Techniques

So far the two-phase and four-phase handshaking protocols are presented. Another dimension of self-timed communication is the use of data encoding. A communication can be carried out using control wires separately of data. This approach is known as bundled-data encoding where it is assumed that by the time request arrives, data is already arrived and is stable. In other words, the delay in data validity indicator (request) wire must be larger than the delay in the data wire. To remove this timing constraints, the data validity indicator signal is included in the data itself resulting a delay-insensitive communication. As already explained, delay-insensitive communication is a necessity for global on-chip communication due to the unavoidable PVT variations and the resulting delay uncertainties in nanometer regime. Most of the existing GALS communication wrappers however have bundled-data encoding interface [26–29].

Hence, there is a need to convert the single-rail data representation to delay-insensitive encoding for global communication. There are many types of delay-insensitive encodings, but the most commonly used in on-chip implementations are dual-rail (1-of-2) and quad-rail (1-of-4) encodings [59]. In a delay-insensitive channel, to transmit N-bit data in parallel it requires $2N + 1$ wires. Only one handshake wire is required since data itself acts as data validity or acceptance indicator depending on the channel type (only acknowledgment wire for push channel and request wire for pull channel). A delay-insensitive channel is shown in Fig. 2.5.

Dual-rail encoding uses two signals to represent each bit of information, and therefore, to transmit N bits of data $2N$ wires are required. Each bit transfer will involve activity in only one of the two wires. As in all delay-insensitive codes, timing information is implicit in the code, that is, it is possible to determine when the entire data word is valid. This is done, for instance, by detecting a level in a four-phase dual-rail data transfer or by detecting a transition in a two-phase transmission on one of the two wires for every bit in the word. A separate wire to convey data readiness is thus not necessary. Transmission of four consecutive bits, 1001 using four and two-phase dual-rail encoding is shown in Figs. 2.6 and 2.7, respectively.

There are other customized dual-rail encodings such as Level-Encoded Dual-Rail (LEDR) encoding, one-phase dual-rail encoding and pulse dual-rail encoding aimed at either minimizing the timing overhead or circuit complexity. LEDR encoding

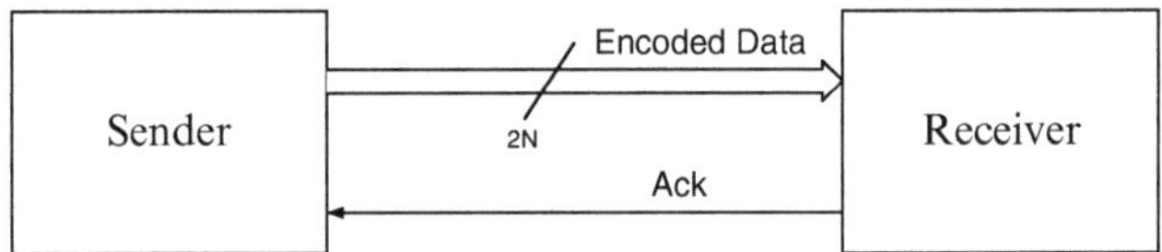

Fig. 2.5 Delay-insensitive push channel

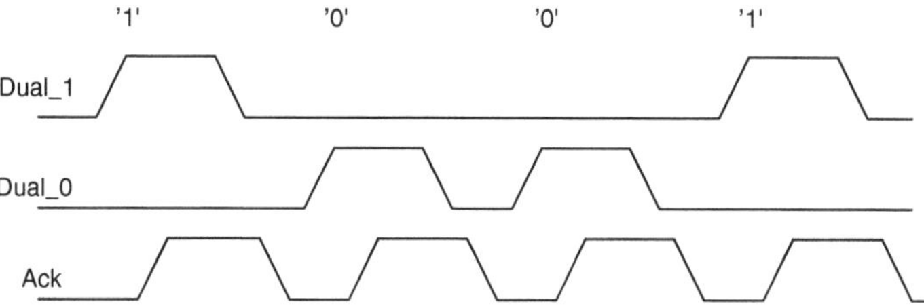

Fig. 2.6 Four-phase dual-rail encoded transmission

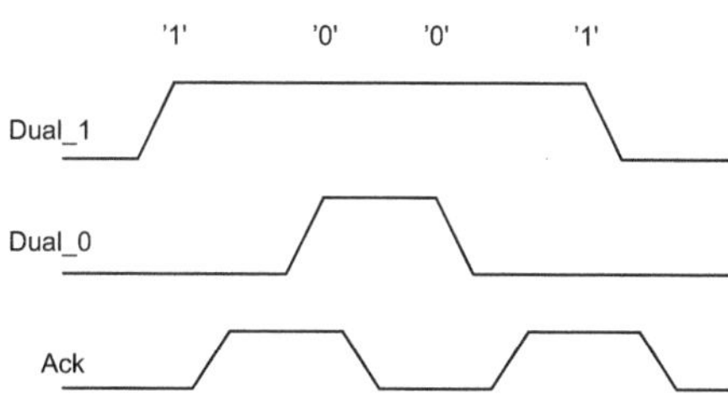

Fig. 2.7 Two-phase dual-rail encoded transmission

is used in one of the on-chip interconnects presented in this book (see Sect. 4.1). Pulse dual-rail encoding has been formulated and used along with wave-pipelining in the serial on-chip link, presented in Chap. 6. It encodes each bit into Pulse and No Pulse (P, NP) pair depending on bundled-data and request signal inputs from the transmitter.

In 1-of-4 data encoding, a group of four wires is used to transmit two bits of information per symbol. A symbol is one of the two-bit codes 00, 01, 10, or 11 and it is transmitted through activity on one of the four wires. Since it is possible to detect the arrival of each symbol at the receiver, 1-of-4 encoding is delay-insensitive. Besides being delay-insensitive, 1-of-4 encoding has more immunity against crosstalk effects as compared to single-rail (bundled-data) encoding, because the likelihood of two adjacent wires switching at the same time is much smaller. In dual-rail encoding, representation of a valid N-bit value requires $2N$ transitions, whereas in 1-of-4 it requires only N transitions. The reduction of transitions in 1-of-4 encoding decreases the dynamic power consumption due to the lower wire capacitance. Transmission of three consecutive symbols using four-phase and two-phase 1-of-4 encoding is illustrated in Figs. 2.8 and 2.9, respectively. Four-phase 1-of-4 encoding with voltage-mode pipelining signaling has been used in [61]. However, in nanometer era where the wire delay dominates over the gate delay the use of four-phase handshaking has significant delay overhead as it requires four communications per transfer. Therefore, the two-phase 1-of-4 data encoding is used in one of the high-performance current sensing on-chip interconnects presented in Chap. 4.

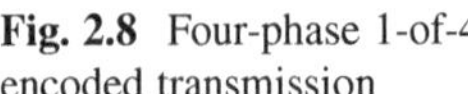

Fig. 2.8 Four-phase 1-of-4 encoded transmission

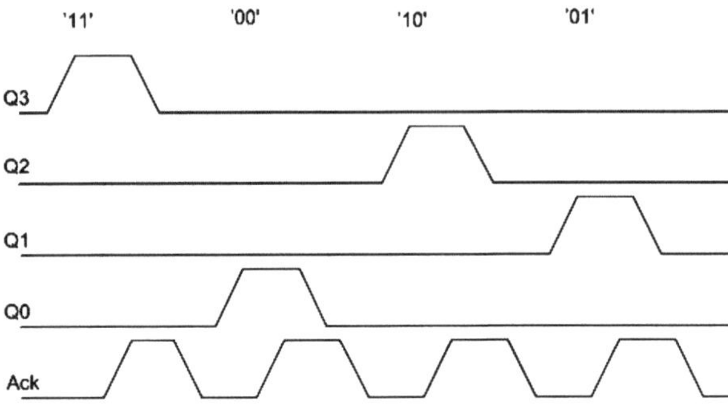

Fig. 2.9 Two-phase 1-of-4 encoded transmission

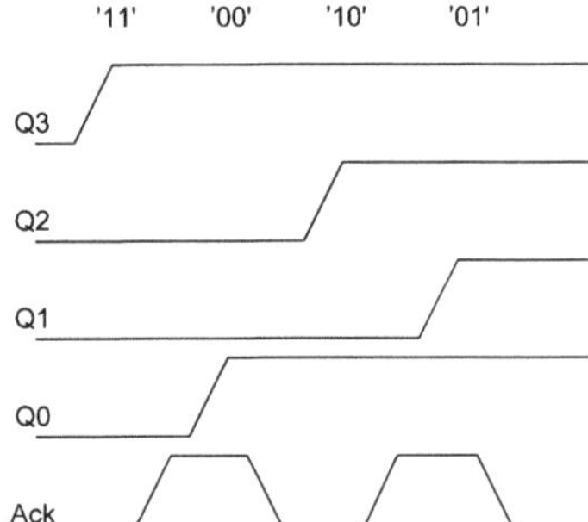

2.3 Data Decoding Techniques

In delay-insensitive data transmission, the receiver has to decode the transmitted encoded data. The complexity of the decoding circuit depends on the chosen encoding. The simpler the decoding logic is, the more attractive is the encoding.

The decoder of four-phase dual-rail encoded channel detects whether one of the dual-rail wires is set to high state or not. The high state indicates a valid data. This detection can be done using an OR gate. The two-phase dual-rail transmission requires more complex decoding logic because it has to detect current transitions on both wires. Furthermore, the decoder has to compare the current transition with the previous transition. Due to this it is not suitable for high-performance communication. In LEDR encoded transmission, data is decoded directly from the state wire using an inverter or buffer to make it full swing. This technique does not require complex decoding logic and the required number of communication actions is two (similar with two-phase dual-rail). Therefore, LEDR is a better alternative over two-phase dual-rail. The pulse dual-rail encoded transmission together with differential pulse signaling, requires no decoding logic. That is, the receiver output provided by the differential amplifier is the transmitted data. This increases the throughput and it is one of the reasons for formulating this type of encoding for the serial link presented in Chap. 6.

The decoding of 1-of-4 encoded data has similar problems as the dual-rail encoding. That is, the decoder needs to sense the voltage levels of the wires, which requires two 2-input OR gates per one 1-of-4 group. The decoding of voltage-mode two-phase 1-of-4 encoded transmission is complex. The decoder consists of XNOR gates which detect the transitions on the wires, NAND gates and a SR latch to decode the data back into the single-rail form. The data decoding in a current sensing 1-of-4 encoded interconnect becomes simpler and faster than voltage-mode one because it does not need to detect transitions and compare with the previous transitions. It consists of current comparators and OR gates (Sect. 4.2).

2.4 Completion Detection Techniques

In synchronous interconnects, the role of the clock is to define points in time where signals are stable and valid. In a self-timed communication, the absence of the clock means that there must be another way to detect when signals are stable and valid. In a delay-insensitive channel, the validity of data is encoded within the data by the transmitter and data validity test (completion detection) is performed by the receiver. The validity test is used to determine that the arrived data is a valid value for the chosen delay-insensitive encoding. In practice, it is also necessary to perform data neutrality test. The implementations of validity and neutrality tests play an important role in the efficiency of a delay-insensitive communication channel.

The completion detection for a four-phase dual-rail (1-of-4) encoded channel is carried out by sensing voltage levels on each pair (one 1-of-4 group) of wires. In a two-phase channel sensing voltage transitions of each pair (group) of wires is required, in this case it requires XOR gates instead of OR gates. Completion detection logic of two-phase dual-rail and 1-of-4 encoded 32-bits channel is shown in Sect. 5.1, Figs. 5.2, and 5.1, respectively. This way of detecting data validity requires logic circuitry whose delay increases drastically when the channel bit width increases, making delay-insensitive interconnects problematic for high performance systems. A fast completion detection technique, where its delay does not increase with transmission bit width, is proposed in Chap. 5 for current sensing interconnects.

2.5 Self-timed Components

In this section, design of self-timed components which are used in the interconnects are discussed briefly. A C-element is a basic building block of self-timed logic. It is a state-holding element, a special kind of latch. When all of its inputs are 0 or 1 the output is set to 0 or 1, respectively. For other input combinations, it preserves its state. Its truth table is shown in Table 2.1 where t and $t-1$ indicate the current and previous values, respectively. Transistor-level implementation of a C-element is shown in Fig. 2.10.

Table 2.1 The truth table of 2-input C-element

a_t	b_t	c_t
0	0	0
0	1	c_{t-1}
1	0	c_{t-1}
1	1	1

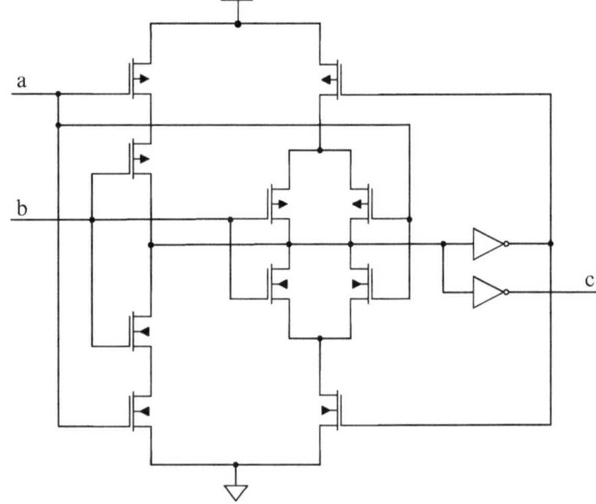

Fig. 2.10 Transistor level implementation of a 2-input C-element

A resettable C-element is a variant of C-element which has a reset input. Its output can be enforced to 0 using the reset input, independently of its other inputs. Its circuit is shown in Fig. 2.11. An active-low resettable C-element has been used in one of the interconnects designed in this book, see Sect. 4.2.3. An upper asymmetric C-element is also a variant of C-element where one of its inputs acts like an active-low reset signal. When all inputs are 1 its output is set to 1 and if the input that acts as active-low reset is *low* the output is set to *low* regardless of the other inputs value. For other input combinations, the C-element preserves its state. A 3-input upper asymmetric C-element has been used in the serializer circuit presented in Sect. 6.2.2. Its CMOS implementation is shown in Fig. 2.12.

2.6 On-Chip Signaling Schemes

The signal transmission systems used in CMOS circuits can be broadly classified into two categories: voltage-mode and current-mode signaling. The important difference between these two transmissions systems lies in the type of the transmitted signal. That is, the signal can be transmitted using voltage or current. Several design options for interconnect signaling exists, for example, single-ended or differential signaling, pulse signaling, and wave-pipelining. A designer has to

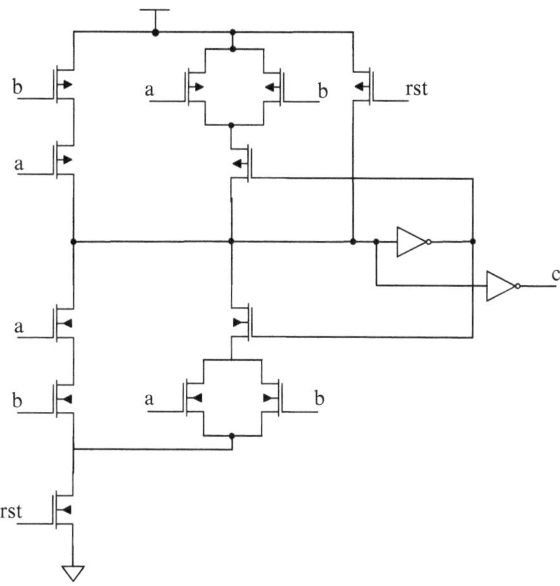

Fig. 2.11 Active-low resettable C-element

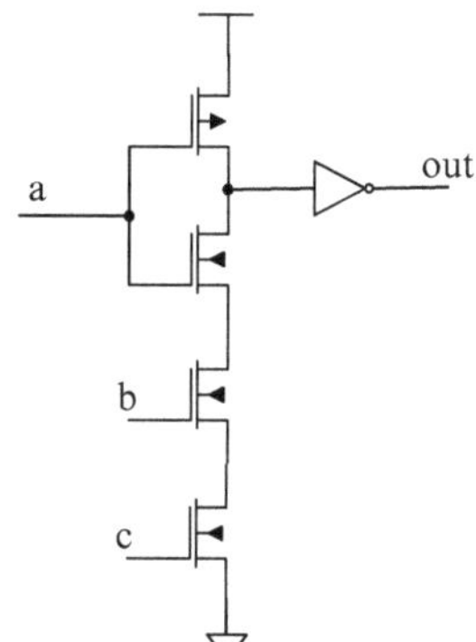

Fig. 2.12 3-input upper asymmetric C-element

choose the optimal signaling scheme and possibly customize it. To do so, there is a tradeoff among latency, throughput, power and area that should be considered. In this section, different signaling techniques that have been designed in order to improve the performance of delay-insensitive on-chip interconnects are discussed. The conventional voltage-mode signaling with repeater insertion and pipelining is also discussed since it has been used as a reference case.

2.6.1 Current-Mode and Current Sensing Signaling

The key to current-mode and current sensing signaling is the low-impedance termination at the receiver which results in reduced signal swings without the need of separate voltage references and increased bandwidth performance. Also this low-impedance termination shifts the dominant pole of the system and leads to a smaller time constant and thus less delay. It is typically implemented by terminating the line with a diode connected transistor. This signaling can operate at a much lower noise margin than the voltage-mode network because the current conveyed to the wires by the current-mode transmitter is well defined and not subject to the effect of either supply voltage fluctuation or ground bouncing. It also operates at a much lower swing as well due to its immunity to power supply noise. All these translate into increased bandwidth performance [30], decreased delay and reduction in dynamic power dissipation and higher noise immunity. The other important feature of current-mode signaling is its reduced delay sensitivity due to process induced variations [31]. For these reasons, current-mode signaling technique becomes a better alternative than voltage-mode for contemporary and future high-speed noise-prone single-chip systems. Current-mode and current sensing signaling have already been proven to provide drastic speed enhancements for on-chip signaling [32–34]. It is also shown theoretically in [32] that current sensing signaling can be three times faster than voltage-mode signaling.

There are three sources of power dissipation in current-mode circuits: static, dynamic, and short-circuit power dissipation. In current-mode signaling static power dissipation is the major component of the total power dissipation that arises from the constant current path from VDD to ground via the termination. Static power dissipation can be minimized using different circuit techniques which reduce leakage currents. Dynamic power is dissipated when the parasitic capacitance of the wire is charged and discharged. Since current-mode signaling operates at low voltage swing dynamic power consumption is not as significant source of power dissipation as in voltage-mode signaling. The third source of power dissipation arises from the finite input signal edge rates that result in short-circuit current. Generally, careful control of input edge rates can minimize the short circuit current component to within 20% of the total dynamic power dissipation [35].

Inspired by the advantages explained above, different signaling techniques based on customization of current-mode or current sensing signaling have been designed and used in all of the presented on-chip interconnects (Chaps. 4 and 6).

Current-mode and current sensing signaling refers to sensing a signal with a low impedance termination at the receive-end which results in a shift or extension in dominant pole position thereby increasing the bandwidth of the line. The difference between these two is their receiver type. That is, in the current-mode signaling the receiver senses the voltage at the end of the wire, compares it with a reference voltage and then amplifies the result. On the other hand, in the current sensing signaling, the receiver senses the current at the end of the wire, compares it with a reference current and finally outputs the result in voltage levels. The current sensing signaling makes the implementation of a delay-insensitive interconnect circuits simpler, especially the data decoding and completion detection circuits as will be discussed in Chap. 4.

2.6.1.1 Binary and Multilevel Current Sensing Signaling

In binary current sensing signaling either there is current I through the wires or there is no current. The receiver compares the wire current with a reference current in order to decode out the transmitted data and also to perform the completion detection test. The LEDR encoded current sensing interconnect, presented in Sect. 4.1, uses a binary current sensing signaling. It uses a diode connected NMOS transistor both as termination load and to mirror the wire current to a current comparator. The current comparator compares the wire current with a reference current.

Using a current comparator with more than one reference current, it is possible to detect more than one current level in the wire. Multilevel current sensing signaling has been proposed for both synchronous and self-timed on-chip interconnects [36–38]. Multilevel current sensing signaling is very attractive for delay-insensitive interconnects because it opens up the possibility to represent each code with a current level. This simplifies the encoding, decoding and completion detection circuits implementation complexity besides minimizing the delay incurred due to decoding and completion detection.

In a delay-insensitive transmission the data validity indicator is the transmitted data itself. The transmission of every new bit needs to be seen in the wire and detected in the receiver. Since two-phase handshake is preferred for long on-chip interconnects either transition in voltage (in case of current-mode signaling) or different current values can be used as data validity indicator. Using transition in current-mode signaling may cause unnecessary power consumption due to the constant current flow in some of the wires which previously made a transition to a high state. In order to save this power, the interconnect presented in Sect. 4.2 allows current flow in the wires only during the respective symbol transmission. If binary current mode signaling is used with this type of power saving transmission scheme, the data validity indicator cannot been seen in the wires when there is consecutive transmission of the same symbol. Thus, in two-phase 1-of-4 encoded current sensing interconnect implementation it becomes possible to differentiate between the consecutive transmission of the same symbols using multilevel currents. The transmitted multilevel current is first detected at the receiver by a detecting circuit based on a current comparator. Then, the encoded voltages are estimated using decoding circuitry.

2.6.1.2 Differential Multilevel Current Sensing Signaling

Differential current sensing signaling has better noise robustness than single-ended signaling. It has been demonstrated that high speed and energy efficient on-chip communication has been achieved using differential current sensing signaling [39–42]. In [43] comparisons between differential current sensing signaling and voltage-mode signaling with optimal repeaters insertion have been performed using 250 nm, 130 nm, 65 nm and 45 nm technologies. Besides its superiority in speed for longer wires, differential current sensing signaling consumes less power than optimal repeaters insertion for activity of 50% and higher and length 4 mm and longer for 130 nm, 65 nm and 45 nm technologies.

In order to get both noise and delay variations robustness, four wires per bit are required if binary current sensing signaling is used. Two wires per bit for the delay-insensitive encoding and two wires per each encoded wire to support differential signaling. This has a much larger area overhead and higher power dissipation, as it requires four wires per bit transmission. By sensing current directions and current values simultaneously both the delay-insensitivity and differential signaling has been achieved with only two wires per bit transmission instead of four. This technique has been implemented in the on-chip interconnect presented in Sect. 4.3. A change in the current level on the wire indicates arrival of new data (delay-insensitivity), while the direction of the current flow reveals the logical value of the transmitted bit. This way of integration leads to more power and area efficient robust communication.

2.6.1.3 Wave-pipelined Differential Pulse Current-Mode Signaling

In pulse signaling only a small portion of the wire is charged during pulse propagation, significantly reducing the amount of capacitance needed to be charged and hence, saving a considerable amount of power over level-based signaling. It has been shown that the use of pulse signaling can save up to 50% of energy compared to level-based signaling with repeater insertion [93]. Furthermore, it has been demonstrated through analytical models that more than 70% power saving could be achieved by combining pulse signaling with wave-pipelining technique without penalties of data throughput [94].

In [34], a prototype 8 Gbps serial link employing pulsed current-mode signaling was manufactured and measured. Sharp current-pulse data transmission was used to modulate transmitter energy to higher frequencies, where the effect of wire inductance is maximized, allowing the on-chip wires to function as transmission lines. In addition to power saving, pulse current-mode signaling mitigates the effect of dispersion due to its return-to-zero signaling scheme in which receiver termination is employed.

The serial link, presented in Chap. 6, employs differential current-mode pulse signaling along with wave-pipelining since this helps to achieve both high-throughput and low-power consumption for global communication. It has combined pulse dual-rail encoding with wave-pipelined differential pulse current-mode signaling, enabling both delay variation and noise robustness.

2.6.2 Voltage-Mode Signaling: Reference

In voltage-mode signaling the voltage has to swing from rail-to-rail over the entire length of the wire. This leads to large dynamic power consumption, larger delay and it also generates power-supply noise [91]. The optimal repeater insertion technique [53, 111] used in voltage-mode signaling, was developed to reduce the wire delay and improve performance of lengthy global interconnections. However, with the increase in number and density of interconnects, the number of repeaters would increase manifold, presenting significant overhead in terms of power and area. Furthermore, as the optimal repeater insertion distance decreases with each technology node due to increased resistive effects of interconnect, the overall improvement in delay can be undermined by the exponential increase in the number of repeaters and associated driver/repeater power dissipation. A higher throughput can be obtained by using pipeline latches instead of repeaters to both amplify the signal and spread the link delay over multiple pipeline stages. This further increases power consumption and area costs compared to the simple repeater approach. Since most of high performance delay-insensitive links use either voltage-mode signaling with optimal repeater insertion or pipelining [61, 112–114], both signaling techniques are employed for the delay-insensitive interconnect presented in Chaps. 4 and 6. Comparison between the conventionally implemented voltage-mode with repeaters/pipelining latches and current sensing delay-insensitive interconnects helps designers to make appropriate decision on which signaling techniques to use for specific circumstances.

2.7 Chapter Summary

In this chapter self-timed delay-insensitive communication techniques and high speed on-chip signaling schemes has been discussed. These are the foundation topics for the next chapters. The techniques presented in this chapter are the enabling factors to achieve delay-insensitivity, higher performance and lower power consumption on-chip communication.

Chapter 3
On-Chip Wire Modeling

A chip is non-functional without wires that connect devices each other. Wires carry signals from one place to another. On-chip wires constitute the lowest level in a hierarchy that spans chip to package-level connections. On-chip wire is not an ideal conductor with zero resistance, capacitance and inductance, but rather it is an unwanted parasitic circuit element. With the increase in circuit performance, complexity, density and levels of integration in nanometer technologies, it is essential to include all parasitic effects during the optimization process. However, this is not a feasible approach due to the large amount of design variables in the optimization process and the overall complexity of the chip. Furthermore, this approach has the disadvantage of not seeing the exact problem, because at a given circuit node, only few dominant parameters affect the overall performance. Thus, designers need to have a clear insight into the parasitic wiring effects, their relative importance and their reduced-order models. Wire parasitics estimation is required to compare different interconnect schemes because interconnect figures of merits (performance, power consumption and noise coupling) [98,99] are functions of wire parasitics. In this book a wire refers to just the metal that interconnects different blocks and the interconnect refers to a wire with its driver and data encoder, load (receiver input impedance) and receiver along with data decoder and completion detector. This chapter discusses briefly methods and basis for estimating wire parasitics and the electrical level modeling of wires.

3.1 Wire Parasitic Estimation and Extraction

Wire parasitic extraction is usually done by representing complex structures as a collection of simple geometric elements and then each parasitic value is combined using superposition or introducing scale factors to obtain the parasitics of the complex structure. There are many commonly used tools which extract the wire parasitics by assuming that the electromagnetic field through interconnects are quasi-static; they

E.E. Nigussie, *Variation Tolerant On-Chip Interconnects*, Analog Circuits and Signal Processing, DOI 10.1007/978-1-4614-0131-5_3,

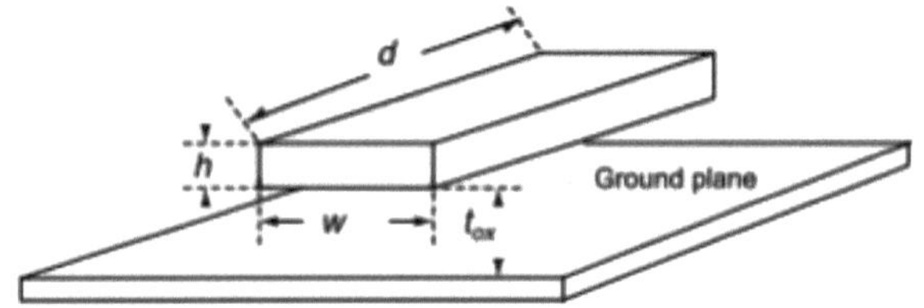

Fig. 3.1 Single microstrip wire

ignore the displacement current in Maxwell equations. With such simplification, electrical fields remain static outside conductors, but magnetic fields retain frequency dependency inside conductors so that the skin effect can be accounted properly. Capacitance and conductance of a structure are determined by electrical fields while resistance and inductance are determined by magnetic fields. In other words, by ignoring the displacement current, magnetic and electrical fields are decoupled in the quasi-static theory. Because of this decoupling, a quasi-static field solver is quicker and can solve much bigger problems than a full-wave solver. For example, FastHenry [45] and FastCap [24] are among the quasi-static field solvers.

The interconnects implemented throughout this book are assumed having a microstrip configuration. A microstrip is a strip of metal over a return ground plane, as shown in Fig. 3.1, where w, h, and d are the wire width, wire height, and wire length, respectively. The t_{ox} is the distance to the underlying ground plane. An electric and magnetic field is created around the microstrip if a driving circuit injects a voltage and current signal, respectively, onto it.

3.1.1 Resistance

The resistance of a wire is the ratio of potential difference of the two ends of a wire to the total current flowing through it:

$$R \equiv \frac{\Phi_{12}}{I} \tag{3.1}$$

where Φ_{12} is the potential difference between the two ends of the wire and I is the current flowing through the wire.

Resistance is dominated by the cross sectional area and the resistivity (inverse of conductivity) of the signal conductor. The DC-resistance, r_{dc} of a microstrip structure, shown in Fig. 3.1, is given by:

$$r_{dc} = \frac{\rho}{h}\frac{d}{w} = R_{square}\frac{d}{w} \tag{3.2}$$

where w, h, ρ are width, thickness and resistivity of the wire, respectively.

Since the thickness is usually a constant for a given technology, it is customary to incorporate it with the resistivity and form a single constant called sheet resistance of the material (R_{square}). At low signal frequencies, Equation 3.2 is sufficient since the entire cross section of the wire carries the current. As the frequency increases, the current density inside is not uniform, but drops away exponentially with depth into the conductor. This phenomenon is called the skin effect since most of the current is now flowing through the skin of the conductor. This leads to current crowding primarily on the surface and the effective cross-section where current flows reduces. As a consequence, wire resistance increases with the frequency. Skin effect is defined as the depth below the surface of the conductor at which the current density decays to 1/e (about 0.37) of the current density at the surface and it is given by:

$$\delta_e = \sqrt{\frac{\rho}{\mu \pi f}} \tag{3.3}$$

Skin effect starts to occur close to the cutoff frequency, fs where $\delta_e \leq 0.3h$ and is fully developed when $\delta_e << h$ (as a guideline $\delta_e \leq 0.1h$) [101]. The obvious and generally accepted term is to get the minimum of width and thickness to obtain the cutoff frequency. For typical on-chip wires, δ_e is found to be equal to $1.5hw/(h+w)$ with relative error less than 5% for $0.25 < h/w < 10$ [100]. There is a widely used empirical formula which describes the frequency dependent behavior of a wire over a ground plane (microstrip structure).

$$R(f) = \begin{cases} r_{dc} & f \leq f_0 \\ r_{dc}\sqrt{\frac{f}{f_0}} & f \geq f_0 \end{cases} \tag{3.4}$$

where

$$f_0 = \frac{\rho}{\mu \pi \delta_e^2}$$

is referred as the break frequency at which this phenomenon begins to dominate. Skin effect decreases the effective cross sectional area that carries the current, which causes resistance to increase. The accurate frequency dependent modeling of wire parameters is usually done considering both resistance and inductance.

Besides the skin effect there are other causes that increase the resistivity of a metal such as metal barrier, surface scattering and temperature. The purpose of the barrier is to prevent the diffusion of copper into the surrounding dielectric. Since the barrier is fabricated from a higher resistivity metal, it is safe to assume that the copper carries all the current but the effective area through which current conducts reduces. As the barrier thickness cannot scale as rapidly as the interconnects, it would increasingly occupy higher fraction of the interconnect cross sectional area while restricting the current flow only to the lower resistivity. The effective resistivity because of barrier is given by [47]

$$\rho_b = \frac{\rho_o}{1 - \frac{A_b}{wt}} \tag{3.5}$$

where ρ_o is the bulk resistivity at a given reference temperature, A_b is the area occupied by the barrier, and w and t are the wire width and thickness, respectively.

Surface scattering has also an effect on wire resistance. Metal resistivity start to increase when the minimum dimension of the metal line becomes comparable to the mean free path of the electrons. This is due to the fact that surface scattering has a significant contribution to the resistivity compared to the contribution from the bulk scattering.

Furthermore, temperature also affects the resistivity of the wire. Conductivity is directly proportional with carrier concentration and mobility of the carriers. Carriers are created by the ionization of atoms within the lattice comprising the solid and the conductors are easily ionized by nature. At a temperature of interest, essentially all the atoms in a conductor are ionized. However, the carriers usually do not move in a straight line when they traverse through a material. This movement is influenced by defects in the lattice, impurities, grain boundaries and fixed ions. As temperature increases, the carriers are more active and suffer more collision, thereby reducing the mobility. In the case of conductors, the mobility is entirely due to ionic scattering and depends on the characteristic of the particular material and can be usually characterized using the conventional relationship [48]:

$$\rho(T) = \rho_o(T_o)[1 + t_{cr}(T - T_o)] \tag{3.6}$$

where $\rho(T)$ is the wire resistivity at any given temperature T, $\rho_o(T_o)$ is the wire resistivity at the reference temperature T_0, t_{cr} is the temperature coefficient of resistance (TCR) of the bulk material. Mathematically, the TCR is slope of $\rho(T)$ vs. T curve normalized to $\rho(T)$, and for the cases where the TCR is nonlinear, a linearized average over a range of temperature may be derived. For bulk Cu, $t_{cr} = 0.39 - 0.43\%^{\text{deg}}\text{C}^{-1}$ at 20^{deg}C [48, 50].

A study on copper wires in 65 nm technology has been carried out by Lu et al. [49] of IBM corporation. They proposed an experimentally validated empirical equation which describes the dependence of wire resistance with surface and grain boundary scattering together with the temperature:

$$\rho_{sg} = \rho_o[1 + t_{cr_bulk}(T - T_o) + \frac{\alpha}{w} + \frac{\beta}{h}] \tag{3.7}$$

where ρ_o is bulk wire resistivity, w and h are wire width and height of the Cu portion, and the model parameters α and β are positive constants, which are functions of a surface scattering coefficient and grain boundary scattering coefficient. α has been extracted for each BOEL levels (that is for each wire thickness h) as $\alpha = a + \frac{b}{h}$. The coefficients are: $a = 0.021\,\mu$m), $\beta = 0.016\,\mu$m, $b = 0.0014\,\mu\text{m}^2$ [49]. They have also found that t_{cr} is equal to be $0.43\%^{\text{deg}}\text{C}^{-1}$ at 20^{deg}C.

3.1.2 Capacitance

When two conducting objects are charged to different electric potentials, an electric field is created between them and a capacitance arises. It always takes some time to build up a voltage between two objects. The capacitance can be seen as the reluctance of voltage to instantaneously increase or decrease in response to an input signal. The capacitance for the single isolated microstrip wire shown in Fig. 3.1, can be approximated by:

$$C = C_{parallel} + C_{fringe} = \frac{w\epsilon_{ox}}{t_{ox}}d + \frac{2\pi\epsilon_{ox}}{ln(2 + 4t_{ox}/h)}d \tag{3.8}$$

where $C_{parallel}$ is the parallel-plate (bottom area-to-substrate) capacitance, C_{fringe} is the fringing (side-wall-to-substrate) capacitance, and ϵ_{ox} is the insulator dielectric constant. This simplification is only useful for estimating rough capacitance values. In reality, a wire is surrounded by a large number of other wires on the same layer and adjacent layers in case of the multilevel structure. Each wire is coupled not only to the grounded substrate, but also to neighboring wires. To model the capacitance in such a complex environment is a non-trivial task and the above equation is not a good model for the capacitance of a wire in such a complicated structure.

In modern ICs, multilevel metal layers are in use and these 3-D interconnects have been simplified to two-dimensional or quasi-three-dimensional structures, based on the layout pattern. If the layers above or below a set of wires in consideration are routed densely, they can be approximated as a ground plane, reducing to a two-dimensional models. Under this condition, capacitive parasitics are scalable functions of wire cross-sectional dimensions. Considering a single wire in multilayer interconnect system, capacitive components can be decomposed into self capacitance (C_s), and mutual capacitance (C_c). In a multilevel wire structure there are two capacitance structures: parallel lines on one plate and parallel lines between two plates. The first structure represents lines without top wiring and the second structure emulates lines with top wiring. In this configurations the presence of adjacent conductors significantly alters the electric field around the central conductor and thus the effect of the wire spacing, s, must be taken into account in the expression of wire capacitances. In [51] self and mutual capacitance formula for a wire has been proposed, which can be used to estimate capacitance of the middle wire using $C_s + 2C_c$. Sakurais mutual capacitance formula is given in equation 3.9 below:

$$C_c = \epsilon\left(0.03\frac{w}{h} + 0.83\frac{t}{h} - 0.07\left(\frac{t}{h}\right)^{0.222}\right)\left(\frac{s}{h}\right)^{-1.34} \tag{3.9}$$

Total capacitance given by $C_s + 2C_c$ is in good agreement with the values predicted by a field solver but individual components are not intended to provide accurate results. In practice, field solver extraction tools are utilized to numerically calculate the parasitic capacitance values. Hence, capacitance values of the wires are extracted using Linpar [46] field solver for the interconnects presented in this book.

3.1.3 Inductance

Inductance is a measure of the distribution of the magnetic field near and inside a current-carrying conductor. This measure is a property of physical layout of the conductor and is also a measure of the ability of the conductor to link magnetic flux or to store magnetic energy. The fundamental equation for inductance is as follows:

$$L = \frac{\oint \vec{B} \cdot d\vec{A}}{I} \tag{3.10}$$

where I is the current, B the magnetic field induced from I, and A is the integration loop. The definition of inductance follows a loop property, the current return path should be known to determine the inductance value. In contemporary interconnect structures the return current is spread all over the range and the exact return path of a current is not known. In these cases, the possible current return paths are the power distribution network and the adjacent wires [102]. The loop formed by the wire and its return path can potentially extend to several hundred micrometers away from the wire under consideration. This vastly complicates the extraction of parasitic inductance of a given wire, as it depends not only on the characteristics of a particular wire, but also on several thousands of other wires. Therefore, in order to find the inductance, the induced current is assumed to return at the infinity. This method was first proposed in [103] and was further introduced for circuit analysis in [104].

A simple approach for inductive parasitic extraction is to use a free space relationship, which relates loop inductance (L) of a wire to its capacitance ($C_{\epsilon_r=1}$) by assuming no dielectric in the medium, the inductance is given by:

$$L = \frac{\epsilon_0 \mu_0}{C_{\epsilon_r=1}} \tag{3.11}$$

This formulation is used in the tool Raphel RC2 [105], which is a two dimensional parasitic extraction tool. Considering the middle conductor in a three parallel conductor system, the self and mutual inductance equations become:

$$L_S = \frac{\epsilon_0 \mu_0}{2} \left(\frac{1}{C_S} + \frac{1}{C_S + 2C_c} \right) \tag{3.12}$$

$$L_m = \frac{\epsilon_0 \mu_0}{2} \left(\frac{1}{C_S} - \frac{1}{C_S + 2C_c} \right) \tag{3.13}$$

where C_s and C_c are self and coupling capacitances, respectively. Unfortunately in an IC, this assumption does not hold and more detailed methods need to be used. For a wire with a finite conductivity, the magnetic flux exists both inside and outside the conductor, subdividing the wire inductance into internal and external components. The internal inductance of a wire is due to the magnetic flux inside the wire and the external inductance is due to a magnetic flux outside the wire (loop or partial inductance is external to the wire). When modeling the internal inductance, high frequency effect of the current distribution has to be considered because of the skin effect. The current distribution inside a conductor also changes with frequency due to the proximity effect, the current tends to concentrate closer to the current return path in order to minimize the inductance. Another effect of frequency on the inductance is due to multi-path current re-distribution. In an IC, there are many possible current return paths, e.g., the power/ground network, nearby signal lines, and the substrate. The distribution of the return current among these possible paths is determined by the impedance of the individual paths. At different frequencies, the relationship among the impedances of different paths will change, as well as the distribution of the return current. The return current is distributed in a way that the total impedance is minimized at a specific frequency. If the frequency dependent effects are very important to consider in a desired frequency range, the cross-sections are subdivided into sections smaller than the skin depth at the maximum frequency of interest. Then, the current distribution in each filament can be regarded as uniform. To calculate the partial inductances of rectangular cross-sectional wires, closed-form equations proposed in [103] are used. In this manner, an inductively coupled RL circuit can be formed for the conductor. By solving currents in this circuit at several points in the frequency domain, the frequency dependent resistance and inductance can be obtained [106]. This technique, which is known as partial element equivalent circuit (PEEC) is the foundation for frequency dependent parasitic extraction tools such as FastHenry [45]. For all the interconnects presented in this book, the inductance and resistance values of the wires were extracted using FastHenry.

3.2 Electrical Level Wire Modeling

In order to analyze the performance and signal integrity of an interconnect, it is necessary to translate the wire layout and technology information such as the width and length of the wire, neighboring line conditions and related dielectrics into electrical parameters. Then these parameters can be combined with other circuit components to evaluate performance. This is achieved through parasitics extraction. Based on the design and technology specifications, a physical line is usually converted into a netlist composed of resistors, capacitors and inductors (if necessary). Due to the technology scaling and increasing operating speeds, accurate modeling of wires has become a necessity. Wires have traditionally been modeled as lumped RC segments but with circuit operation frequency on the rise, this model lacks the required accuracy to model a high-performance interconnect.

The fundamental electrical behavior of a metal wire can be fully determined using Maxwell's equations (Equations 3.14–3.18) in conjunction with the rule of charge conservation (Equation 3.18).

$$\nabla \cdot D = \rho \tag{3.14}$$

$$\nabla \times E = \frac{\partial B}{\partial t} \tag{3.15}$$

$$\nabla \cdot B = 0 \tag{3.16}$$

$$\nabla \times H = J + \frac{\partial D}{\partial t} \tag{3.17}$$

$$\nabla \cdot J + \frac{\partial \rho}{\partial t} = 0 \tag{3.18}$$

Since solving these equations requires a huge amount of computation, they are usually simplified depending on the range of frequencies and wire lengths of interest. As already discussed in the above section, the behavior of a wire is frequency dependent. At DC it behaves as a resistor, causing both losses in the voltage supply (IR drop) and static power consumption (IR^2). Wire activities are also affected by the interaction between electric and magnetic fields when operating in AC range. In current IC designs, quasi-static assumption is usually applicable since the signal frequency is relatively low and the wire length is much shorter than the wavelength of the signal. For instance at 10 GHz, the wavelength is about 17 cm for k = 3.0 dielectrics. As explained in Sect. 3.1, under the quasi-static assumption the electric and magnetic field can be decoupled. Thus, wire capacitance and inductance can be defined and extracted independently and the resulting wire is represented by an RC or RLC equivalent circuit. To solve the electrical response, the wire is assumed to be uniform, and therefore the Maxwell's equations can be reduced to the telegraph equation (transmission line theory) as follows:

$$\frac{\partial^2 V}{\partial x^2} = RC\frac{\partial V}{\partial t} + LC\frac{\partial^2 V}{\partial t^2} \tag{3.19}$$

where x is the length dimension, t is the time and V is the voltage. A dimensionless ratio of the physical length of a wire to the signal wavelength, $\frac{\ell}{\lambda}$, is referred as the electrical length. This ratio is used to determine whether to model the wire using a lumped or distributed model. A wire is considered to be electrically short if the electrical length is less than unity. These electrically short wires belong to the classical circuit analysis and it is quite safe to approximate the entire line as a lumped RC or RLC segment because the signal level along the entire length of the wire is almost constant. A rule of thumb to determine whether a wire can be represented by a lumped circuit or not is to test its length against the following criterion [107]:

$$length \leq \lambda/20 \tag{3.20}$$

where λ is the signal wavelength. Since the frequency spectrum that a digital signal contains is more closely related to its rise time ($\frac{1}{3.14t_r}$) than to the signal frequency itself, λ should be estimated from the rise time of the signal. Transmission line modeling needs to be applied when the time of flight (time required for a signal to travel round trip from the driver to the end of a line) across the wire becomes comparable to the signal rise time. A transmission line can be thought as a large number of lumped segments in series so that they represent the distributed nature of the wire.

The importance of modeling inductive effects in wires is increasing because of faster rise times and longer wires. Wide wires used in upper metal layers can be

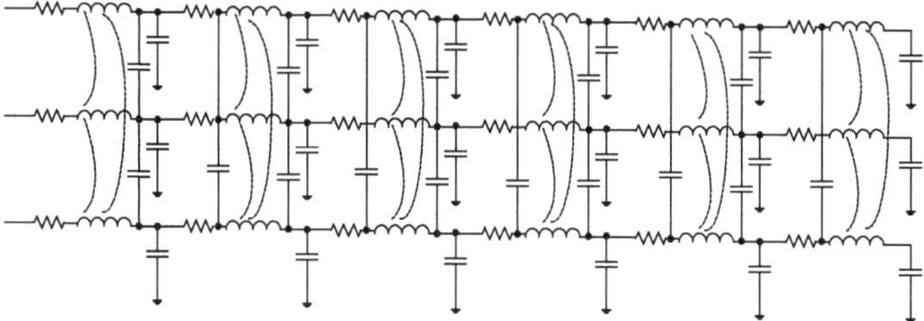

Fig. 3.2 Distributed RLC wire model with coupling

especially susceptible to inductive effects due to their low resistance [96]. Wires should be modeled as RLC lines if they satisfy the following two conditions [96, 107]: input signal rise time is smaller than the time of flight and the time of flight is greater than the Elmore delay of an RC line. The latter criteria describes a situation where wire resistance is considerably smaller than line characteristic impedance. By combining these two criteria, the following condition is obtained:

$$\frac{t_r}{2\sqrt{LC}} < length < \frac{2}{R}\sqrt{\frac{L}{C}} \tag{3.21}$$

where R, L, and C are the resistance, inductance and capacitance per unit length, respectively. In case the constraint on the left-hand side of Equation 3.21 is larger than the right-hand side, $t_r > 4L/R$, the input signal is not fast enough and the inductance effect can be ignored regardless of the wire length.

Since the interconnects designed in this book are targeted for high-performance signaling over global wires, all wires are modeled using a distributed RLC model by considering the inductance effect, as shown in Fig. 3.2. In order to accurately consider crosstalk noise effects, both capacitive and inductive coupling between all wires was also included.

3.3 Chapter Summary

Wires are not ideal as drawn in schematic diagrams but a parasitic element which exhibits undesired effects and degrades the performance of electronic systems. These non-idealities are usually captured by computing the electromagnetic behavior of a wire using field solvers. In this chapter, wire parasitic extraction and electrical level wire modeling are discussed briefly. Parasitic extraction requires expensive simulators and a lot of computational time. The standard approach to reduce this complexity is to partition the problem into a set of geometry dependent parasitics and solving a discrete electrical network made up of parasitic elements.

The basic requirement in this partitioning is to have both efficiency in simulation and the required level of accuracy. Electrical models of wires takes different forms depending on accuracy and computational complexity. Major questions in selecting an electrical model is when to consider inductance and the frequency dependency in the models.

Chapter 4
Design of Delay-Insensitive Current Sensing Interconnects

Unlike synchronous design style which uses a globally distributed clock signal to indicate moments of stability of the data, asynchronous circuits exchange information using handshakes to explicitly indicate the validity and acceptance of data. Depending on the type of handshaking, data encoding, channel type, and data-validity schemes there are a number of alternative communication protocols. As already discussed in Chap. 2, two-phase handshaking is preferred for global on-chip communication since it reduces the number of transitions and avoids the requirement of a spacer between consecutive data symbols. This saves communication time and energy of the system. The most common asynchronous data encoding in GALS design is bundled-data (single-rail) encoding which uses N lines to represent N-bit information and two additional handshake lines indicating data validity and acceptance. Since this encoding has a timing constraint between control (data validity) and data lines, communication through a long on-chip interconnect becomes sensitive to delay variations. Therefore, converting bundled-data encoding to delay variation insensitive encoding is necessary for global on-chip interconnects where delay variations are unavoidable. The general block diagram of conversion between bundled-data and delay-insensitive encoding is shown in Fig. 4.1. The conversion between the two encodings requires a data encoder at the transmitter side and a data decoder as well as a completion detector at the receiver side.

In this chapter design and analysis of three delay-insensitive current sensing on-chip interconnects are presented. Their performance and power consumption are analyzed and compared with conventional delay-insensitive on-chip interconnects. The design and analysis of each interconnect are presented in separate sections. The performance, energy and area of these three interconnects are compared to each other using the same technology and wiring models in Chap. 7.

This chapter is organized as follows. In Sect. 4.1, design of an on-chip interconnect which uses LEDR encoding and current sensing signaling is presented. Analysis of its performance and power consumption along with two dual-rail encoded reference interconnects are also discussed. The design and simulation results of a 1-of-4 encoded multilevel current sensing interconnect is presented

E.E. Nigussie, *Variation Tolerant On-Chip Interconnects*, Analog Circuits and Signal Processing, DOI 10.1007/978-1-4614-0131-5_4,

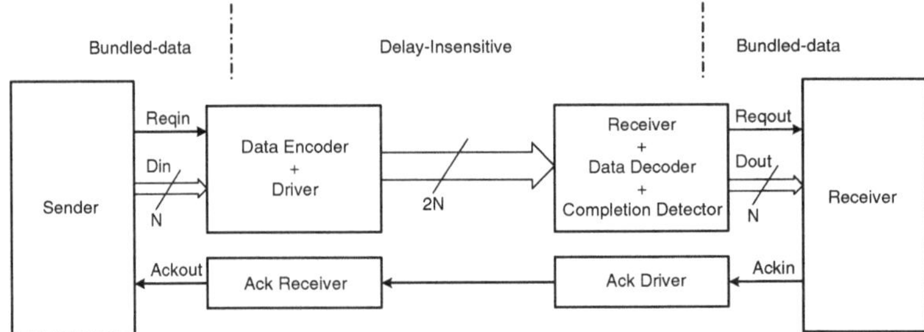

Fig. 4.1 Bundled-data ⇔ delay-insensitive conversion

Fig. 4.2 Conventional two-phase dual-rail encoder

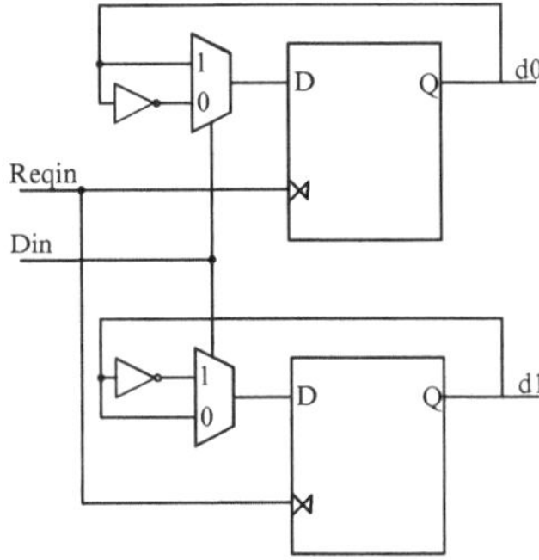

in Sect. 4.2. Its performance and power efficiency has been compared with two 1-of-4 encoded voltage-mode interconnects. In Sect. 4.3, area and power efficient two-phase dual-rail encoded differential current sensing interconnect is presented. The summary of this chapter is presented in the last section.

4.1 Level-Encoded Dual-Rail Current Sensing Interconnect

LEDR encoding is among the preferred encoding schemes for global on-chip communication, because it needs no resetting transitions that consume time and power. Its completion detection and decoding circuitry is faster and much simpler than that of two-phase dual-rail encoding since detection is level based rather than transition based. The conventional two-phase dual-rail protocol has more complex and slower decoding and completion detection circuitry compared to LEDR. In the two-phase protocol, if the transmitted data has the value 0 there is a transition on one wire and a transition on the other wire if 1 is transmitted. To detect completion and decode the data, the current and previous state on both wires need to be detected. This makes the circuit relatively complex and slow. The gate level implementations of the encoder, decoder and completion detector of two-phase dual-rail encoded transmission are shown in Figs. 4.2 and 4.3.

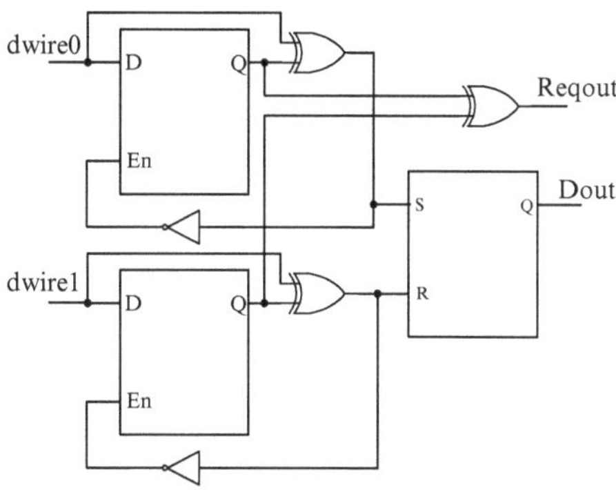

Fig. 4.3 Conventional two-phase dual-rail decoder and completion detector

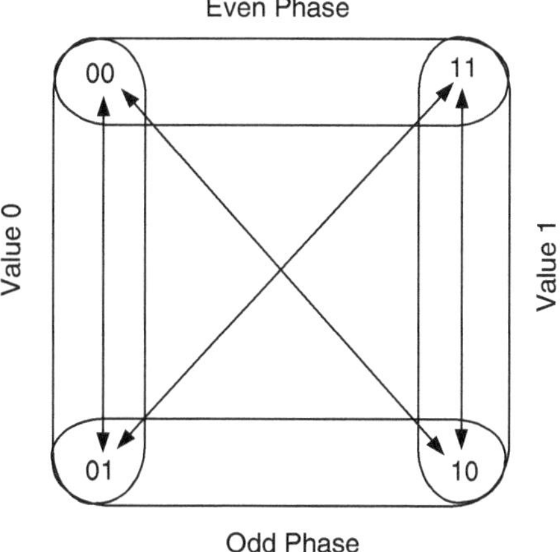

Fig. 4.4 Transitions between LEDR code

In LEDR one data bit is encoded into a 2-bit codeword as follows. A data sequence D(i) of bits is encoded into a sequence DS(i) and DP(i) of state and phase bits, respectively.

$$DS(i) = D(i), \quad \forall\, i$$

$$Given\ DP(0),$$

$$\mathbf{if}(DS(i+1) = DS(i))\ \mathbf{then}$$

$$DP(i+1) = \neg DP(i),$$

$$\mathbf{else}$$

$$DP(i+1) = DP(i)$$

As each codeword has a phase, even or odd, it is possible to differentiate between two consecutive same bit transmissions. Figure 4.4 shows the four possible codewords organized into overlapping groups of value and phase. The arrows illustrate the allowed transitions between the codewords.

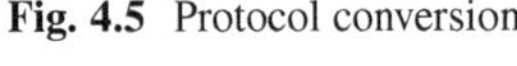

Fig. 4.5 Protocol conversion

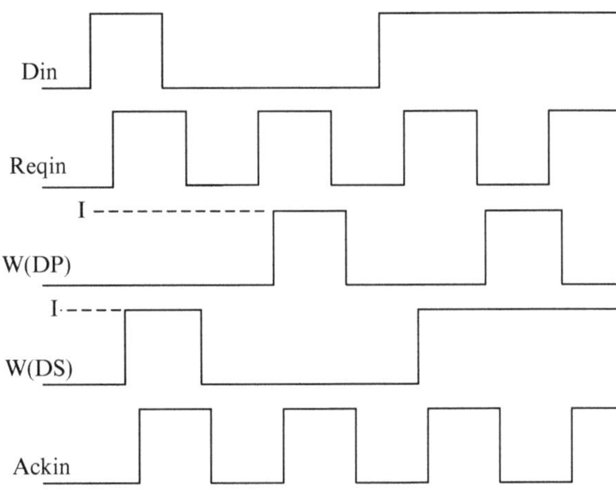

4.1.1 Data Encoder and Driver

The encoder takes the request and data bit in the voltage-mode bundled-data form and converts this information into current-mode LEDR signaling. The conversion from the two-phase voltage mode to the LEDR current mode is shown in Fig. 4.5 at the protocol level. As shown in Fig. 4.6, the outputs of the double-edge-triggered flipflops 1 and 2 (*DFF1* and *DFF2*) control the current flow through the phase and state wires (*DP* and *DS*) by serving as gate voltages of the transistors *Mn2* and *Mn4*. Considering the data phase wire *W(DP)*, the transistors *Mp1* and *Mn1* generate the source current. The transistor *Mp2* is used to mirror the generated current *I* from the current source and drive this current through *W(DP)* when *Mn2* is '*ON*'. The same principle applies to the state wire *W(DS)*.

4.1.2 Receiver, Decoder and Completion Detector

At the receiver side, the current comparator circuit, as depicted in Fig. 4.6, is composed of the diode-connected input NMOS transistor *Mn6*, the NMOS transistor *Mn7* connected to replicate this input current, the threshold current generating pair of transistors *Mn5* and *Mp3*, and the PMOS transistor *Mp4* that replicates the threshold current. In addition to serving as an input transistor, *Mn6* acts also as a termination load. The drains of the PMOS replicating transistor *Mp4* and NMOS replicating transistor *Mn7* are connected to generate the comparator circuit's output voltage *V(DP)*. The comparator provides a logical *high* output voltage when the input current *I(DP)* is less than the threshold current and a logical *low* output voltage when the input current *I(DP)* is greater than the threshold current.

As shown in Fig. 4.6, the decoder takes as input the output of the state wire's current comparator, *V(DS)*, and reconstructs the data sent by the transmitter. Unlike

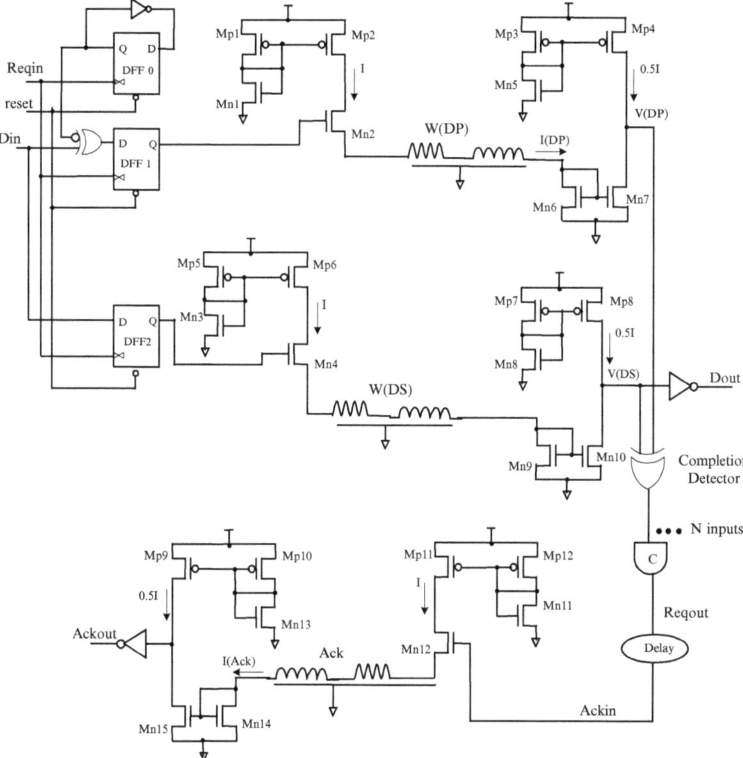

Fig. 4.6 LEDR encoded current sensing on-chip interconnect

conventional two-phase transmission which needs to detect both wires' current and previous states (Fig. 4.3), here the decoder needs sensing voltage level of the state (*W(DS)*) wire using only an inverter. The completion detection is carried out using 2-input XOR gate, the outputs of the two current comparators are the inputs for the completion detector. The completion detection circuit is also simpler and faster, just one XOR gate per each LEDR group is needed. For *N* bit transmission, completion detection is carried out using *N* 2-input XOR gates connected to an *N*-input C-element. The output of the C-element acts as the bundled-data request signal (*Reqout*) passed to the receiving module.

4.1.3 Acknowledgment Transmission

The acknowledgment signal transmission circuitry is also shown in Fig. 4.6. The voltage-mode bundled-data acknowledge signal (*Ackin*), sent by the receiving module, is converted into a current-mode signal *Ack* during transmission and back into a voltage-mode signal (*Ackout*) at the transmitter side. As can be seen from

Fig. 4.6, the signaling circuits are basically equivalent to the ones used for data transmission. When *Ackin* is *high*, current *I* flows through the *Ack* wire causing an up-going transition on the signal *Ackout* at the transmitter side. When *Ackin* goes *low*, the current is switched off and *Ackout* eventually returns to zero.

4.1.4 Simulation Results and Analysis

Latency, throughput and average total power consumption are considered as main parameters to evaluate the presented LEDR on-chip interconnect (*LEDRCm*). Also the performance and power consumption of two reference interconnects are analyzed. One of the reference interconnects uses LEDR encoding along with voltage-mode signaling with repeaters (*LEDRVm*). The other one uses two-phase dual-rail encoding and voltage-mode signaling with repeaters(*TPDRVm*). This helps to determine the performance improvement and power overhead due to the use of current-mode signaling along with LEDR encoding over *LEDRVm* and *TPDRVm*. During simulations a transmission line model of the wires was assumed by using 20 distributed RLC sections. Metal 4 of a 130 nm CMOS technology with minimum metal width, spacing and pitch was used to model the transmission line. The resistance and inductance matrices of the interconnect structure were extracted using FastHenry [45], while the capacitance matrices were extracted using Linpar [46]. The interconnect circuitry was designed and simulated using Cadence Analog Spectre with 130 nm CMOS technology from STMicroelectronics. The supply voltage was 1.2 V.

Here forward latency is defined as the delay from a transition on the bundled-data request signal (*Reqin*) at the transmitter side to the corresponding transition on the bundled-data request signal (*Reqout*) at the receiver side (see Fig. 4.6). Reverse latency is defined as the delay from a transition on the bundled-data acknowledgment signal (*Ackin*) at the receiver side to the corresponding transition on the bundled-data acknowledgment signal (*Ackout*) at the sender side. The change in forward and reverse latency when the wire length is varied from 1 to 11 mm is shown in Figs. 4.7 and 4.8 for *LEDRCm*, *LEDRVm* and *TPDRVm* interconnects. The *LEDRCm* interconnect latencies are much smaller than *LEDRVm* and *TPDRVm* interconnects for longer wires. For example, at 7 mm communication distance, the forward latency of *LEDRCm* is about two-thirds of *LEDRVm* and half of the *TPDRVm* latencies. The forward latency of *TPDRVm* interconnect is higher than the two LEDR encoded interconnects, showing the impact of its complex encoding/decoding and completion detection logics. The latency difference between *LEDRCm* and *LEDRVm* interconnects shows the use of current sensing signaling in enhancing the performance of delay-insensitive interconnects especially at global wire lengths. The forward latency of 9 mm long *LEDRCm* interconnect is only 46% of that of the *LEDRVm* interconnect's latency with the same communication distance.

As in the latency, the *LEDRCm* interconnect throughput is higher than that of the *LEDRVm* and *TPDRVm* interconnects. At 7 mm long communication distance,

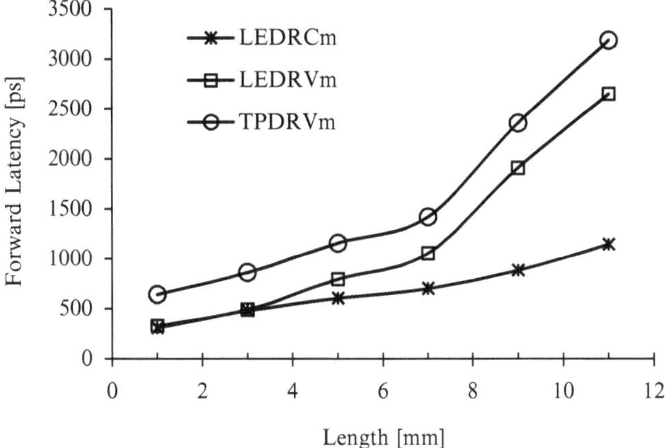

Fig. 4.7 Forward latency of *LEDRCm*, *LEDRVm* and *TPDRVm*

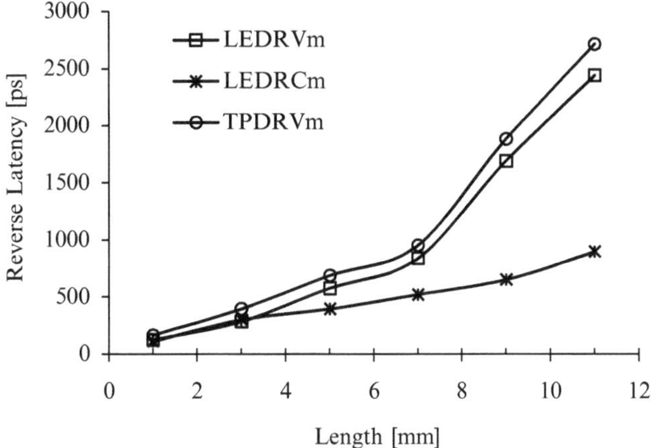

Fig. 4.8 Backward latency of *LEDRCm*, *LEDRVm* and *TPDRVm*

throughput of *LEDRCm* interconnect is 1.55 and 1.94 times higher than that of *LEDRVm* and *TPDRVm*, respectively. Its throughput is not dropping as fast as the other two interconnects with the increase in the communication distance, this can be seen from Fig. 4.9. The *LEDRCm* interconnect achieves 1.005 Gbps throughput per one dual-rail group (two data transmission wires +1 acknowledgment wire) at 5 mm wire length without using repeaters or pipelining. The main reason for the higher throughput is the use of current sensing signaling. If a number of these dual-rail groups concatenate in parallel, throughput increases linearly albeit the completion detection (data arrival and stability check) for all groups deviate the linear increase slightly (see Chap. 6).

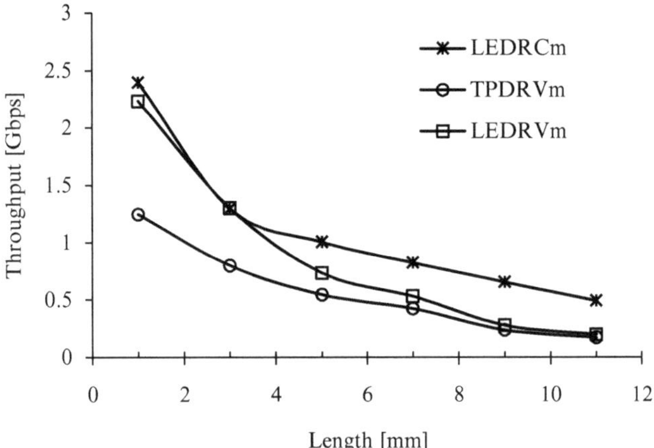

Fig. 4.9 Throughput of *LEDRCm*, *LEDRVm* and *TPDRVm*

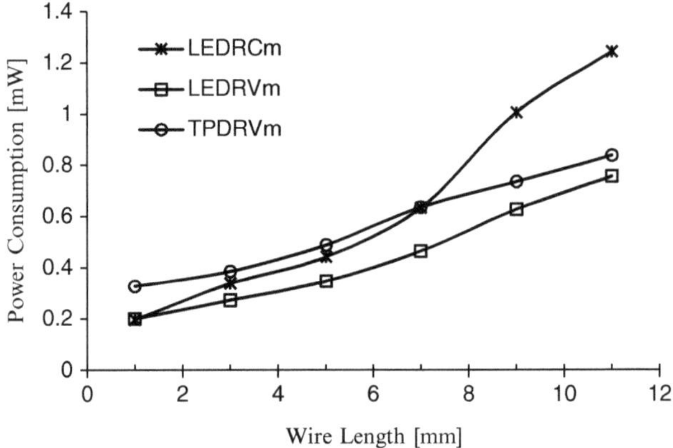

Fig. 4.10 Power consumption of *LEDRCm*, *LEDRVm* and *TPDRVm*

The average total power consumption of the one-bit *LEDRCm* interconnect varies from 195μW to 1242μW when wire length is varied from 1 to 11 mm. Its power consumption is higher than that of the two reference interconnects especially with long wires as shown in Fig. 4.10. However, it is more power efficient for longer communication distances as can be seen from the energy per bit diagrams in Fig. 4.11. At global wire lengths, that is, starting from 5 mm, it dissipates least energy per bit compared to the other two, and *TPDRVm* dissipates the most.

To summarize the throughput improvement and energy savings of *LEDRCm* interconnect over the reference interconnects (*LEDRVm* and *TPDRVm*), Tables 4.1 and 4.2 are presented. Table 4.1 shows the advantage of current sensing signaling by comparing *LEDRCm* and *LEDRVm* interconnects. The benefit of LEDR encoding along with current sensing signaling over conventional two-phase dual-rail encoded

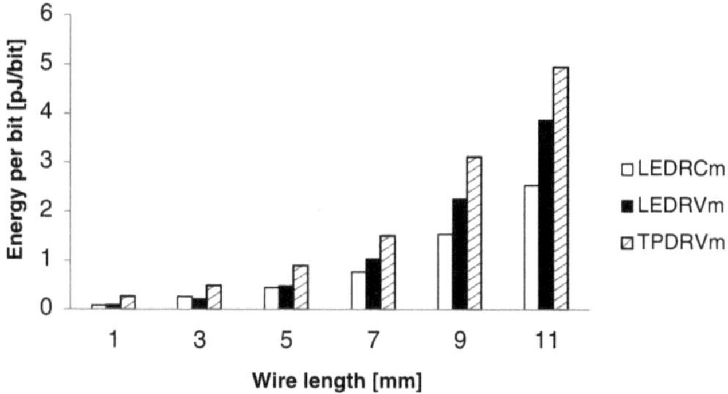

Fig. 4.11 Energy per bit dissipation of *LEDRCm*, *LEDRVm* and *TPDRVm*

Table 4.1 Comparing *LEDRCm* and *LEDRVm* interconnects

Wire length [mm]	Gain in throughput	Energy savings (%)
5	1.37*X*	7.24
7	1.82*X*	25.36
9	2.34*X*	31.65
11	2.5*X*	34.28

Table 4.2 Comparing *LEDRCm* and *TPDRVm* interconnects

Wire length [mm]	Gain in throughput	Energy savings (%)
5	1.85*X*	51.14
7	1.95*X*	48.91
9	2.76*X*	50.51
11	2.89*X*	48.78

voltage-mode interconnect is demonstrated in Table 4.2. *LEDRCm* interconnect gains almost double throughput and 50% energy savings compared to *TPDRVm*.

The simulation waveforms of the one-bit *LEDRCm* interconnect are shown in Fig. 4.12. As can be seen, there is a change in current only in one wire per data transfer either on *W(DS)* or *W(DP)*.

4.1.5 Effect of Crosstalk on Timing

Since *LEDRCm* is a delay-insensitive interconnect, crosstalk induced signal propagation delay variations can cause only performance penalty, do not affect its reliability. In this section the performance penalty caused by crosstalk is examined. In this analysis, 4-bit parallel data transmission is considered. This requires 8 parallel physical wires since delay-insensitive encoding is used. The wires are modeled as transmission lines with both capacitive and inductive coupling between each other. Minimum wire separation distance of 210 nm is used with minimum

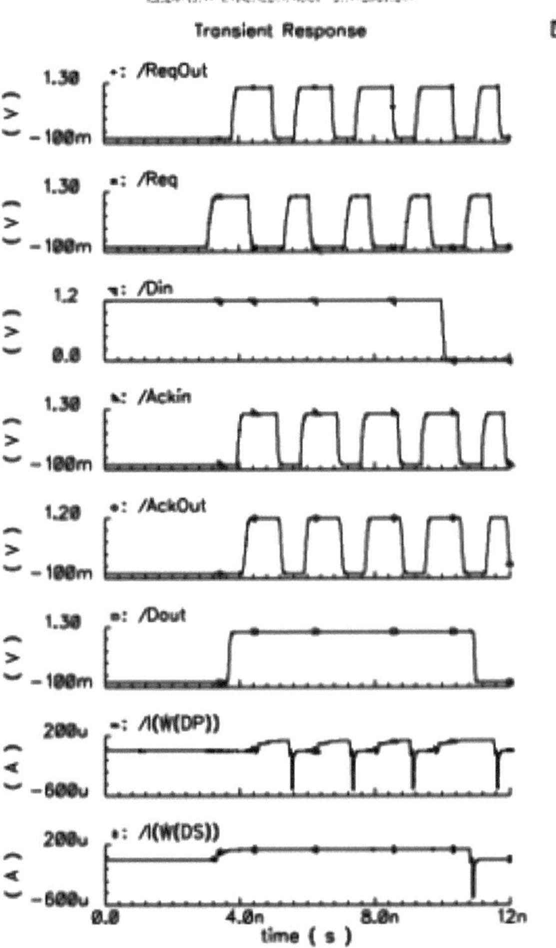

Fig. 4.12 Simulation waveforms of *LEDRCm* interconnect

global pitch specified in 130 nm technology and 1.2 V supply voltage. The worst-case switching pattern was defined by assuming that capacitive coupling dominates inductive coupling, which is the most usual case in on-chip parallel wires. The effect of crosstalk on performance of *LEDRCm* interconnect is compared with bundled-data voltage mode (*BundledVm*) interconnect. Because the worst-case switching pattern of *LEDRVm* is the same as *LEDRCm*. Furthermore, from crosstalk effect on timing perspective, bundled-data encoded interconnect can represent synchronous transmission. The delay variation percentage of the *LEDRCm* interconnect due to worst-case crosstalk is less than one-third of that of a bundled-data voltage mode (*BundledVm*) one, as shown in Table 4.3. This is because in LEDR encoded data transmission only the state wires (*W(DS)*) make transitions when there is a switching in the input data. The phase wires (*W(DP)*) are quiet. At most the victim wire has one nearest aggressor.

Table 4.3 Effect of crosstalk in *LEDRCm* and *BundledVm*

Interconnect	Worst-case switching	% Delay variations
BundledVm	↑↑↓↑ − − −−	+141
LEDRCm	− ↑ − ↑ − ↓ − ↑	+42

4.2 1-of-4 Encoded Current Sensing Interconnect

In 1-of-4 data encoding, a group of four wires is used to transmit two bits of information per symbol. A symbol is one of the two-bit codes 00, 01, 10, or 11 and it is transmitted through activity on one of the four wires. Since it is possible to detect the arrival of each symbol at the receiver, 1-of-4 encoding is delay-insensitive, as are all the 1-of-N codes [52]. Besides being delay-insensitive, 1-of-4 encoding has more immunity against crosstalk effects when compared to bundled-data encoding, because the likelihood of two adjacent wires switching at the same time is one-eighth times smaller. Furthermore, dynamic power consumption due to wire capacitance is smaller for the 1-of-4 code than for the simpler 1-of-2 (dual-rail) code. This is because the 1-of-4 code conveys two bits of information using only a single transition, while the 1-of-2 code requires two transitions for two bits of information.

In this section, implementation of a novel high-performance link based on multilevel current sensing signaling and delay-insensitive two-phase 1-of-4 encoding is presented. Current sensing signaling reduces communication latency of global wires significantly compared to voltage-mode signaling, making it possible to achieve high throughput without pipelining and/or using repeaters. Performance of the proposed multilevel current-mode interconnect is analyzed and compared with two reference voltage-mode interconnects.

The 1-of-4 Encoded Multilevel Current Sensing (*PMCm*) scheme converts two-phase bundled-data voltage-mode signaling into pulsed 1-of-4 multilevel current sensing signaling at the transmitter side. At the receiver side, delay-insensitive current sensing signaling is turned back into bundled-data voltage-mode communication. The *PMCm* scheme is logically equivalent to a 1-of-4 encoded voltage-mode scheme, the difference is that information is presented as current pulse rather than voltage transitions, as shown in Table 4.4. Hence, one of the four data wires draws current to indicate the presence of a new two-bit data symbol. Similarly, an acknowledgment is signaled as current on the acknowledgment wire. As explained in Chap. 2, such a current sensing implementation is inherently much faster and more immune against power supply noise and delay variations compared to a voltage-mode implementation. The communication protocol is shown in Fig. 4.13 (from the receivers perspective) and the signaling circuits are depicted in Figs. 4.14 and 4.15. The advantage of this interconnect implementation is that high throughput and low latency can be achieved without using area and power hungry pipelining or repeaters.

The multilevel and pulsed nature of the *PMCm* scheme can be seen in Fig. 4.13. The current detected at the receiver has three different values: 0, I_1 and I_2.

Table 4.4 Encoding and wire current of *PMCm*

Bundled-data		PMCm			
$D_1 D_0$	Reqin	$I(WQ_3)$	$I(WQ_2)$	$I(WQ_1)$	$I(WQ_0)$
00	$0 \to 1$	0	0	0	I_2
	$1 \to 0$	0	0	0	I_1
01	$0 \to 1$	0	0	I_2	0
	$1 \to 0$	0	0	I_1	0
10	$0 \to 1$	0	I_2	0	0
	$1 \to 0$	0	I_1	0	0
11	$0 \to 1$	I_2	0	0	0
	$1 \to 0$	I_1	0	0	0

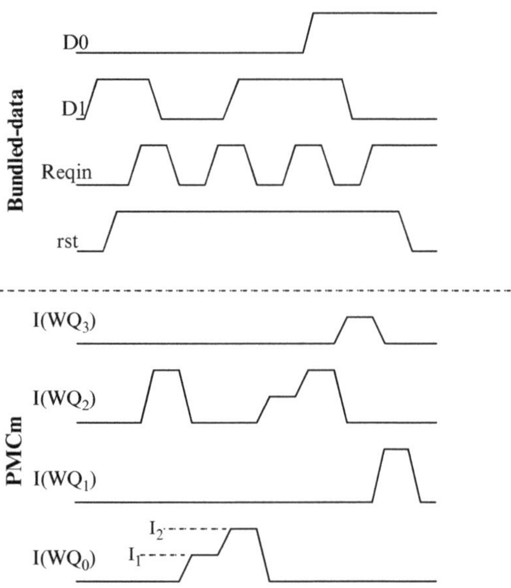

Fig. 4.13 Communication protocol of *PMCm*

The values I_1 and I_2 are used when the voltage-mode request signal *Reqin* at the transmitter side is *low* and *high*, respectively, reflecting the adopted two-phase communication protocol. The value 0, in turn, means that there is no symbol on a wire. It is used as the initial value of the data wires and for switching off current on a wire when the 2-bit symbol to be transmitted changes, making current on a wire pulse shaped. This feature reduces the overall power consumption of the current-mode interconnect. The values of I_1 and I_2 are determined by considering the speed, power consumption, and noise margin of the interconnect. In the following consecutive sections, the implementations of the encoder, decoder and completion detector are separately discussed.

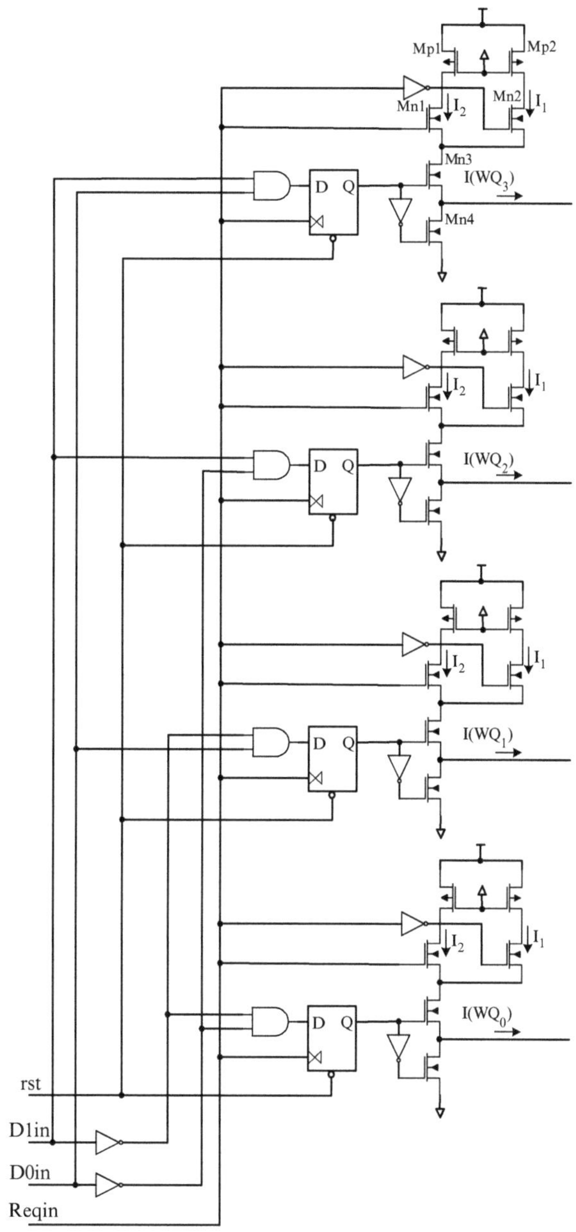

Fig. 4.14 Encoder and driver of *PMCm*

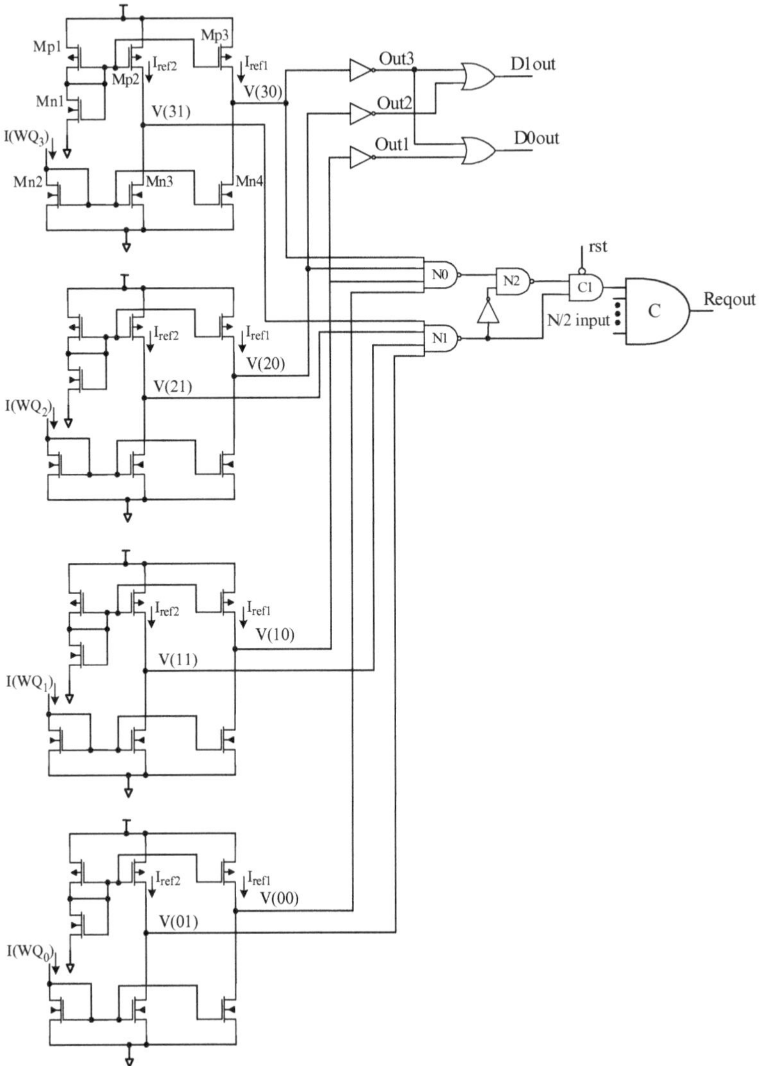

Fig. 4.15 Decoder and completion detector of *PMCm*

4.2.1 Encoder and Driver

The encoder takes the request and two data bits in the voltage-mode bundled-data form and converts this information into multilevel current sensing 1-of-4 signaling. The double-edge triggered flip-flops shown in Fig. 4.14 are used to sample the value of the 2-bit data symbol at each transition of the two-phase request signal *Reqin*. For instance, consider the encoder circuit of the wire $Q3$. Depending on the value of the signal *Reqin*, either transistor *Mn1* or *Mn2* conducts making either current I_1 or I_2 to flow through the wire $Q3$ when the symbol 11 has arrived from the sender module. To prevent the line from drawing current continuously, the transistor *Mn4* is used to ground the line when other than the symbol 11 is sent. The reset signal *rst* is controlled by the transmitting module. When a data burst is about to begin, *rst* is set to *high* enabling the sampling flip-flops. When the burst has been completed, *rst* is initialized back to *low*, meaning that all the data wires become grounded. This is necessary to prevent the data wires of the link from drawing current (consuming power) during possibly long idle periods between bursts. In nanometer scale technologies process variation effects are one of the major concerns. The driver output currents may vary from their expected values due to process variation effects. In order to minimize this variation, transistors *Mp1* and *Mp2* which operate in the linear region form a resistive path from the supply voltage to *Mn1* and *Mn2* which in turn keeps the switching threshold of *Mn1* and *Mn2* transistors constant.

4.2.2 Receiver

At the receiver side, consider the current comparator circuit of $Q3$, as depicted in Fig. 4.15. It is composed of the diode-connected input NMOS transistor *Mn2*, the NMOS transistors *Mn3* and *Mn4* connected to replicate this input current, the reference or threshold current generating pair of transistors *Mn1* and *Mp1*, and the PMOS transistors *Mp2* and *Mp3* that replicate the threshold current. In addition to serving as an input transistor, *Mn2* acts also as a termination load. The drains of the PMOS reference current replicating transistors and line current replicating NMOS transistors are connected together to generate the comparator circuit's output voltages, $V(30)$ and $V(31)$. This comparator provides a logical high output voltage when its input current $I(Q3)$ is less than the threshold current and a logical low output voltage when the input current $I(Q3)$ is greater than the threshold current. Here the current comparator compares current on the wire $Q3$ with two different threshold currents, I_{ref1} and I_{ref2}, in order to distinguish the three current levels. To be more specific,

$$\begin{aligned}
&\mathit{If}(I(Q3) < I_{ref1})\ \mathit{then}\\
&\quad (V(30) = 1) \wedge (V(31) = 1) \quad //(\mathit{initial\ state})\\
&\mathit{If}(I_{ref1} < I(Q3) < I_{ref2})\ \mathit{then}\\
&\quad (V(30) = 0) \wedge (V(31) = 1)\\
&\mathit{If}(I(Q3) > I_{ref2})\ \mathit{then}\\
&\quad (V(30) = 0) \wedge (V(31) = 0)
\end{aligned}$$

In nanometer scale technologies, the line and reference currents at the input of the receiver may vary from the nominal value due to supply voltage, process and temperature variation effects. In Chap. 8 different techniques are developed to ensure reliability of the communication by restoring the current levels within the desired margins at power start-up.

4.2.3 Decoder and Completion Detector

As shown in Fig. 4.15, the data decoder, composed of three inverters and two OR gates, needs as inputs the outputs of the current comparators of the wires $Q3$, $Q2$, and $Q1$ to reconstruct the two bits (*D1out, D0out*) sent from the transmitter module. Only the comparator outputs of the threshold current I_{ref1} (i.e., $V(10)$, $V(20)$, and $V(30)$) are needed for this purpose. Formally, the logic is as follows:

$$\begin{aligned}
&\mathit{If}(V(30) = 0) \wedge (V(20) = 1) \wedge (V(10) = 1)\ \mathit{then}\\
&\quad (D1out = 1) \wedge (D0out = 1)\\
&\mathit{If}(V(30) = 1) \wedge (V(20) = 0) \wedge (V(10) = 1)\ \mathit{then}\\
&\quad (D1out = 1) \wedge (D0out = 0)\\
&\mathit{If}(V(30) = 1) \wedge (V(20) = 1) \wedge (V(10) = 0)\ \mathit{then}\\
&\quad (D1out = 0) \wedge (D0out = 1)\\
&\mathit{If}(V(30) = 1) \wedge (V(20) = 1) \wedge (V(10) = 1)\ \mathit{then}\\
&\quad (D1out = 0) \wedge (D0out = 0)
\end{aligned}$$

The completion detector reads all current comparator outputs as illustrated in Fig. 4.15. For each 4-wire block, the completion detection circuit includes two 4-input NAND gates (*N0* and *N1*), a 2-input NAND gate (*N2*), and a resettable 2-input C-element (*C1*). To produce the receiver-side request signal *Reqout*, the completion signals of the *N*/2 4-wire blocks are combined with an *N*/2-input C-element, where *N* is the bit-width of the transmitted data. The completion detection process is started by sensing the current values on the four wires. In this pulsed implementation of 1-of-4 encoding, current flows only in one of the four wires. Current through the wire becomes I_1 or I_2 when the transmitter-side request signal *Reqin* is *low* or *high*, respectively. Hence, if the input current of the comparator is greater than the threshold I_{ref2}, then the output of the C-element *C1* and subsequently the receiver-side request signal *Reqout* go *high*. Correspondingly, if the comparator input current is between the thresholds I_{ref1} and I_{ref2}, the output of *C1* and the signal *Reqout* go *low*. The completion detection logic uses as inputs the current comparator outputs $V(30)$ and $V(31)$ of $Q3$, $V(20)$ and $V(21)$ of $Q2$, $V(10)$ and $V(11)$ of $Q1$, and $V(00)$ and $V(01)$ of $Q0$. For instance, consider again the receipt of the symbol 11

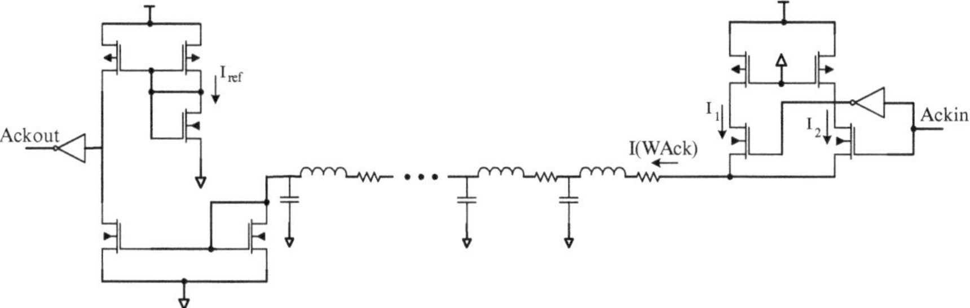

Fig. 4.16 Acknowledgment transmission of *PMCm*

through the wire $Q3$. Assuming that the transmitter-side request signal *Reqin* is *high*, the current on the wire $Q3$ is I_2. Consequently, the comparator outputs $V(30)$ and $V(31)$ become *low*, and all the other comparator outputs remain high since no current flows through the wires $Q2$, $Q1$, and $Q0$. This makes the outputs of the NAND gates *N1* and *N2* high, causing an up-going transition on the output of the C-element *C1*. Formally, the completion detection logic for the symbol 11 is as follows (The output of a gate X is denoted by O(X)):

$$
\begin{aligned}
&(V(30) = 0) \wedge (V(31) = 0) \quad \text{(current is } I_2) \\
&\Rightarrow (O(N0) = 1) \wedge (O(N1) = 1) \\
&\Rightarrow (O(N2) = 1) \\
&\Rightarrow (O(C1) = 1) \\
&(V(30) = 0) \wedge (V(31) = 1) \quad \text{(current is } I_1) \\
&\Rightarrow (O(N0) = 1) \wedge (O(N1) = 0) \\
&\Rightarrow (O(N2) = 0) \\
&\Rightarrow (O(C1) = 0)
\end{aligned}
$$

4.2.4 *Acknowledgment Transmission*

The voltage-mode bundled-data acknowledge signal (*Ackin*), sent by the receiver module, is converted into a current-mode signal during transmission and back into a voltage-mode signal (*Ackout*) at the transmitter side. In this interconnect design, transmission of the acknowledgment signal also uses multilevel current sensing signaling. The driver and receiver circuits of this transmission along with distributed RLC model of the Acknowledgment wire is shown in Fig. 4.16. The current through the acknowledgment wire becomes I_1 or I_2 when the acknowledgment signal from the receiving module is *low* or *high*, respectively. The receiver uses a current comparator circuit to detect the value of the current through the acknowledgment wire and output the result in voltage form. An inverter is used to amplify the comparator's output to full-swing.

4.2.5 Reference Voltage-Mode Interconnects

This reference interconnect also uses a two-phase protocol and 1-of-4 encoding, the difference being that it is implemented using voltage-mode signaling. In the *TPVm* scheme one of the four wires makes a transition to indicate the presence of a new two-bit symbol. When this new symbol arrives to the receiving module, the receiver accepts the symbol and sends an acknowledgment to the sender module by changing the state of the acknowledge signal. Since voltage-mode signaling is used, the voltage on the interconnect swings from rail-to-rail over its entire length. This leads to large dynamic power consumption, large delay, and generation of power-supply noise. The usual approach to improve the performance of a voltage-mode interconnect is to insert repeaters or pipeline latches. Inserting repeaters decreases the signal propagation delay at the cost of increasing power consumption and chip area. A higher throughput can be obtained by using pipeline latches instead of repeaters to both amplify the signal and spread the link delay over multiple pipeline stages. This further increases power consumption and area costs compared to the simple repeater approach. Here both schemes are considered for the reference *TPVm* interconnect. The pipelined and repeater-based implementations are called *TPVmP* and *TPVmRep*, respectively. In the *TPVmP* implementation pipeline stages are inserted in every 2 mm along the link wire. This is based on the assumption that the typical distance between two neighboring (adjacent) routers in the on-chip mesh structure is 2 mm [44] and that the local link length can be considered an upper limit for pipeline-free signal transmission as in [78]. In the *TPVmRep* implementation optimal repeater insertion is used for both data and acknowledgment transmission. The required optimal number of repeaters and optimal size of the repeater are calculated using equation (36) of [53]. Using this equation the required number of optimal repeaters becomes $2.22 * L$ and the optimum size of the repeater becomes $76.5*$ minimum size inverter, where L is the wire length in mm.

The straightforward gate level implementations of the encoder which converts the two-phase bundled-data input to the delay-insensitive two-phase 1-of-4 protocol, the pipeline stage, and the decoder and completion detector which converts the delay-insensitive code back to the two-phase bundled-data form at the receiver side are shown in Figs. 4.17–4.19, respectively. The encoder consists of NOR gates which generate the select inputs for the multiplexers depending on the two-bit input codes, double-edge triggered flip-flops which are used to sample the symbol value at both edges of the request signal, and multiplexers each of which allows transition on the corresponding flip-flop output only when the appropriate input symbol is present. The decoder and completion detector circuit consists of XNOR gates which detect

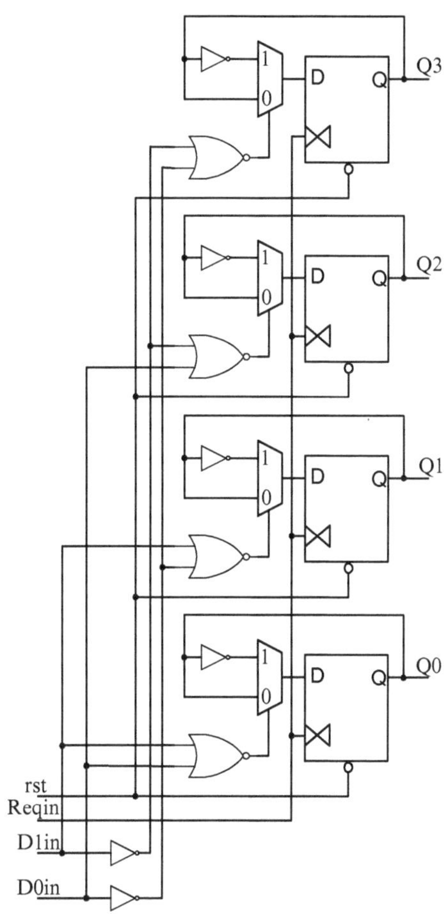

Fig. 4.17 Encoder of TPVm

the transitions on the wires, NAND gates and a SR latch to decode the data back into the bundled-data form, and a four-input XOR gate together with an *N*/2-input C-element for detecting completion. An inverter is used as both driver and receiver for the transmission of the two-phase acknowledgment signal between the pipeline stages in the *TPVmP* implementation, as shown in Fig. 4.18.

4.2.6 *Simulation Results and Analysis*

Simulation of *PMCm* and the two reference voltage-mode interconnects (*TPVmP* and *TPVmRep*) was carried out in Cadence Analog Spectre and Hspice using 130 nm technology from STMicroelectronics, and the supply voltage was set to 1.2 V. Since high-performance signaling over long wires is considered, the wires were modeled using a distributed RLC model of metal 4. In order to accurately model crosstalk

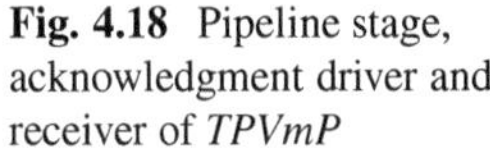

Fig. 4.18 Pipeline stage, acknowledgment driver and receiver of *TPVmP*

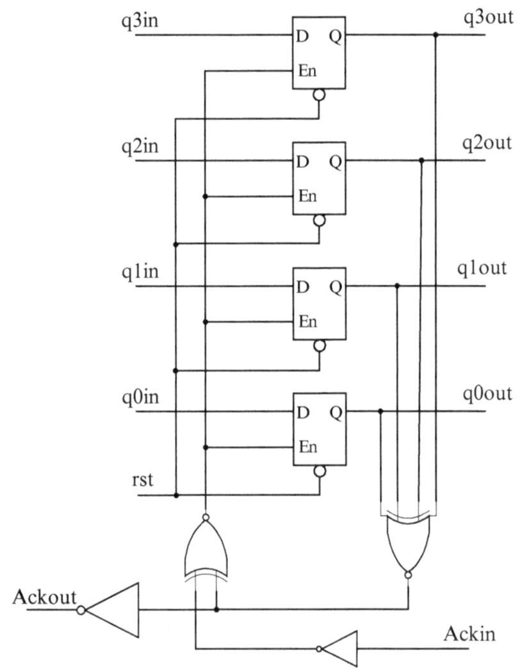

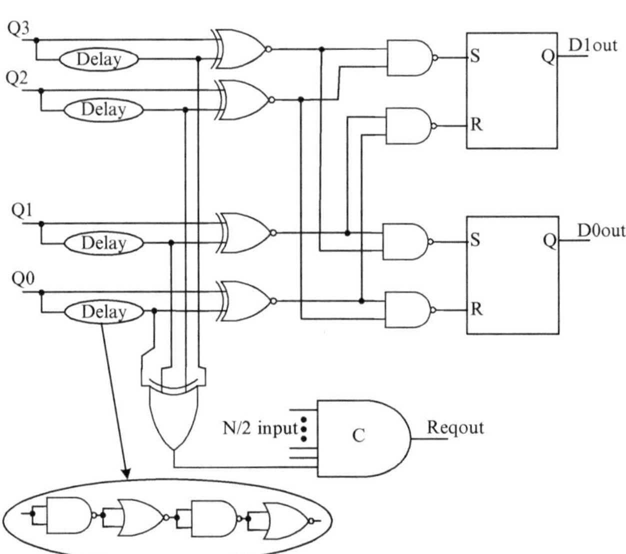

Fig. 4.19 Decoder and completion detector of *TPVm*

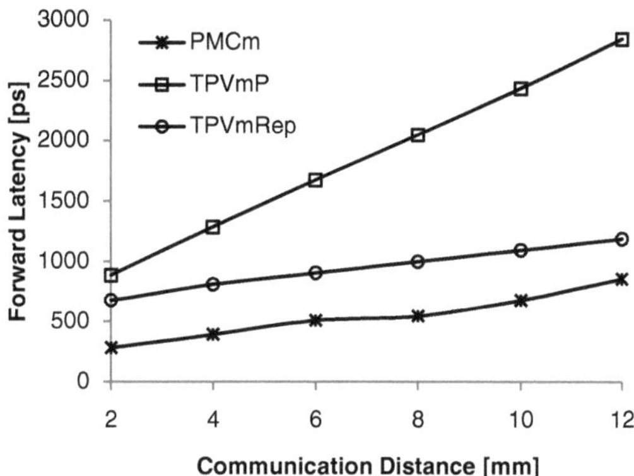

Fig. 4.20 Forward latency of 1-of-4 encoded interconnects

noise, both capacitive and inductive coupling between all wires was included. The bus consisted of eight parallel wires. The RLC values of the wires were extracted using FastHenry [45] and Linpar [46] field solvers as in LEDR encoded interconnect in Sect. 4.1. The wire length was varied in the simulations from 2 mm to 12 mm.

4.2.6.1 Performance Analysis

Latency and throughput are considered the main parameters to analyze the performance of the multilevel current sensing on-chip interconnect along with the two reference voltage-mode interconnects. In the first reference interconnect, *TPVmP*, pipeline stages are inserted every 2 mm assuming that the local wire length (between neighbor routers in a network) is 2 mm. This improves the throughput at the expense of increased forward latency, power consumption and chip area. In the second reference interconnect, *TPVmRep*, optimal size repeaters are inserted at optimal distances.

Here forward latency is defined as the delay from a transition on the bundled-data request signal (*Reqin*) at the transmitter side to the corresponding transition on the bundled-data request signal (*Reqout*) at the receiver side (see Fig. 4.1). In other words, the time required for one flit to traverse from the sending router to the receiving router. The change in the forward latency of the three interconnects when wire length is varied from 2 mm to 12 mm is shown in Fig. 4.20. Since the *PMCm* interconnect uses current sensing signaling, its forward latency is much smaller than the latency of the two reference interconnects. The *PMCms* forward latency was less than one third of *TPVmP*'s latency for all simulated wire lengths. The forward latency of the pipelined voltage-mode interconnect was greater than 1.5 times *TPVmRep*'s latency for 4 mm and longer communication distances.

The throughput of *PMCm*, along with the two reference interconnects, is shown in Fig. 4.21. The throughput of *PMCm* varied from 5.102 Gbps to 1.602 Gbps when the wire length was varied from 2 to 12 mm. At global communication distance of 8 mm, the throughput of *PMCm* was 1.53 and 1.88 times the throughput of *TPVmP*

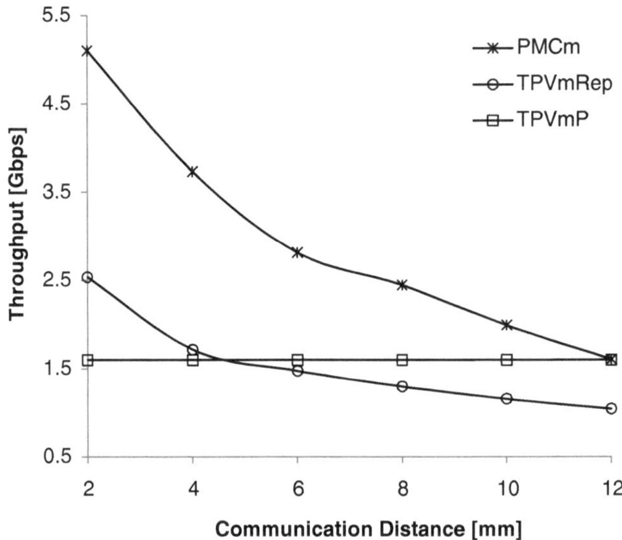

Fig. 4.21 Throughput of 1-of-4 encoded interconnects

and *TPVmRep*, respectively. In the case of the reference interconnects, *TPVmP* achieved a throughput of 1.597 Gbps while the throughput of *TPVmRep* varied from 2.534 Gbps to 1.041 Gbps when the wire length was varied from 2 to 12 mm. The reported latency and throughput values are for one group of 1-of-4 encoding.

The *PMCm* interconnect is a better alternative than *TPVmP* or *TPVmRep* to realize high-performance long-range links. In addition to achieving high performance, *PMCm* circuitry takes a smaller chip area compared to voltage-mode reference interconnects, *TPVm*. This is because the complexity and required chip area of the encoder and decoder of both *TPVm* and *PMCm* interconnects are almost the same. However, the number of required pipeline stages and the number of repeaters increase with wire length, which leads to increase in layout complexity and required area.

4.2.6.2 Power Analysis

The average total power consumption for 2-bit data transfer on the proposed current sensing and the two reference interconnects when communication distance was varied from 2 to 12 mm is shown in Fig. 4.22. *PMCm* has consumed 38% or more power than that of *TPVmP* at all wire lengths. The power consumption of *TPVmP* increases at a faster rate with wire length compared to *PMCm* due to the increase in the number of pipeline stages. As a result the power consumption difference between these two interconnects decreases at global wire lengths. *PMCm*'s power consumption was 10 to 36% lower than that of *TPVmRep* starting from 6 mm wire

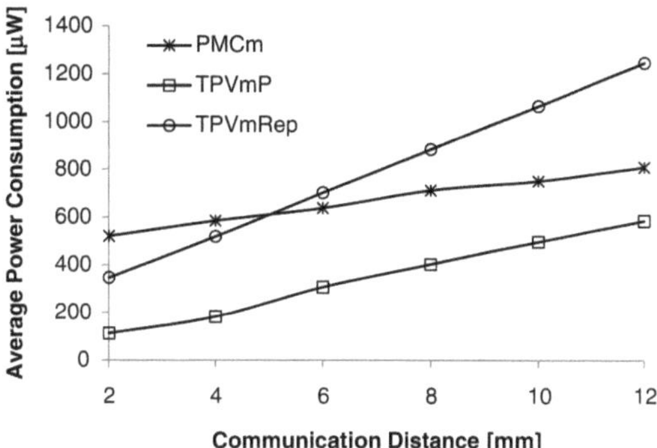

Fig. 4.22 Power consumption of 1-of-4 encoded interconnects

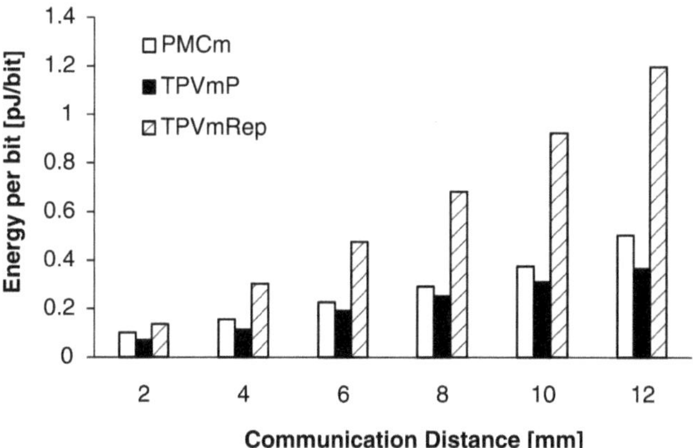

Fig. 4.23 Energy per bit dissipation of 1-of-4 encoded interconnects

length, this is because of the increase in the number of repeaters inserted at global lengths of the wire. The power dissipated by the *TPVmRep* interconnect was higher than 2 times *TPVmP*'s consumption for all wire lengths.

The energy per bit of the interconnects is shown in Fig. 4.23. The energy per bit of *PMCm* was 26 to 58% less than that of *TPVmRep* and 15 to 37% larger than the *TPVmP*'s energy dissipation. *TPVmP* and *TPVmRep* dissipate least and highest energy, respectively at all wire lengths.

4.2.6.3 Noise Analysis

The impact of crosstalk noise on latency and throughput was also studied. In this analysis, 4-bit parallel data transfer was assumed. This requires 9 (8 parallel data transmissions + 1 acknowledgment) physical wires since 1-of-4 encoding is used. The acknowledgment wire was designed as having shielding from the parallel data transmission wires, to counteract the coupling effect. The wires were

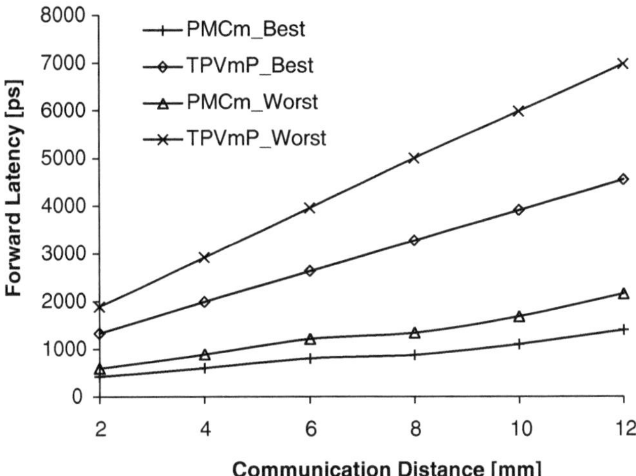

Fig. 4.24 Crosstalk effect in latency of *PMCm* and *TPVmP*

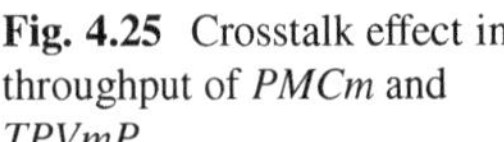

Fig. 4.25 Crosstalk effect in throughput of *PMCm* and *TPVmP*

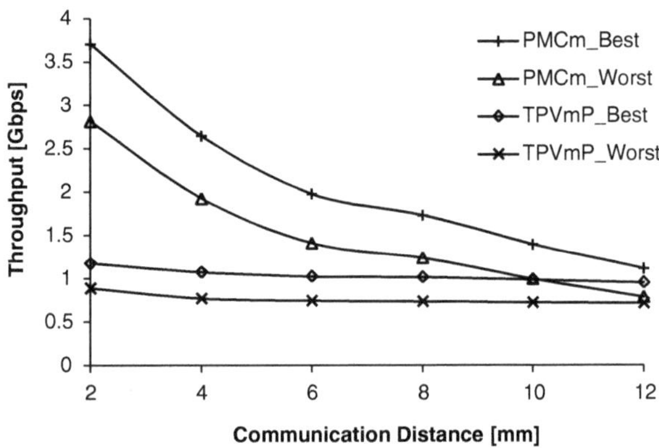

modeled as transmission lines which have both capacitive and inductive coupling between each other. During this analysis, minimum wire separation distance with minimum global pitch specified in 130 nm technology and 1.2 V supply voltage were used. The delay variation due to both capacitive and inductive coupling was simulated by considering the worst-case and best-case switching patterns. These switching patterns depend on the RLC values of the wire. In the simulation setup it is assumed that the capacitive coupling dominates the inductive coupling which is the most usual case in on-chip parallel wires. The effect of crosstalk on latency and throughput when the wire length was varied from 2 mm to 12 mm is shown in Figs. 4.24 and 4.25, respectively.

During best-case and worst-case switching, the latency variation of *TPVmP* was slightly less than that of *PMCm*. For example, at a wire length of 8 mm, the increase in latency due to best-case switching from the crosstalk free latency of *TPVmP* and *PMCm* was 59.8% and 62.3%, respectively. In worst-case switching, the *TPVmP* and *PMCm* latency variations were 144% and 147%, respectively, at the same wire length. In fact, these percentage values are rather large because in the nominal case shown in Fig. 4.20 the considered capacitive loads were only to ground. In other words, the nominal case capacitive loads do not consider the loading effect of the coupling capacitances. The decrease in throughput due to crosstalk was greater for *TPVmP* than for *PMCm*, specially at long wire length. For example at 12 mm wire length, the throughput of *TPVmP* was decreased by 38% while the *PMCm* was only by 30% (Fig. 4.25).

The simulation waveforms of PMCm interconnect are shown in Fig. 4.26. As can be seen, there is current only in one of the four wires at each symbol transmission time. The three current levels in the wire can also be seen from the waveforms.

4.3 Dual-Rail Encoded Differential Current Sensing Interconnect

As already discussed in Chap. 1, global on-chip interconnects get slower with technology scaling and dissipate more power. At the same time signal integrity issues become challenging due to crosstalk, PVT variations and noise. PVT variations cause the signal propagation delay to be uncertain, which in turn affects the performance and reliability of the interconnect significantly. It has been demonstrated that high speed, energy efficient and better noise immunity can be achieved using differential current-mode signaling (see Sect. 2.6.1.2). In addition, reliable on-chip communication in the presence of delay variations is possible through the use of self-timed delay-insensitive data transfer. Thus, integrating differential signaling with delay-insensitive data transfer enables high-performance as well as robustness towards both noise and delay variations. However, integrating these two techniques has considerable area and power overhead as it requires four wires per bit (two for delay-insensitive encoding and two for differential signaling). In this section, a high-performance on-chip interconnect based on novel area and power efficient integration of delay-insensitive data transfer and differential current sensing signaling is presented.

The proposed Dual-rail encoded differential current sensing interconnect (*Dualdiff*) implements both delay-insensitive and differential signaling schemes with only two wires per bit by using a novel encoding and differential current sensing. This leads to a smaller area and smaller power consumption. Its encoding technique and circuits are discussed in the next sections. As in the other interconnects presented in this chapter, both its input and output signals are assumed to be in the two-phase bundled-data encoded form.

4.3.1 Encoding and Its Implementation

In conventional delay-insensitive data encoding transmission of *N*-bit of data requires *2N* wires. The doubling of the wire count compared to bundled-data

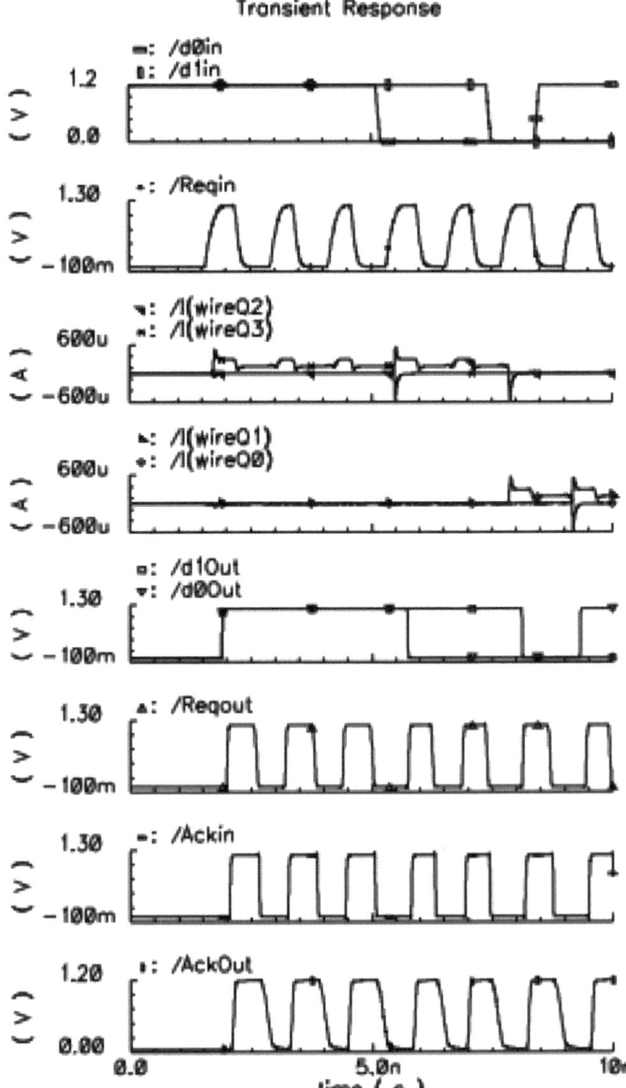

Fig. 4.26 Simulation waveforms of *PMCm*

encoding has a significant effect on the wiring area and routing complexity. Delay-insensitive data transmission using *N* instead of *2N* wires for four-phase [37] and two-phase [36] handshaking has been proposed using single-ended multilevel current-mode signaling. Both of these works use different current levels to encode the data and data validity indicator together.

In the novel encoding technique considered here, current directions and current values are used simultaneously to get both delay-insensitivity and differential

Table 4.5 Encoding protocol of *Dualdiff* interconnect

Din	Reqin	En	Ind1	Ind2	Ind3	Ind4
0	$1 \rightarrow 0$	1	1	0	0	0
0	$0 \rightarrow 1$	1	0	0	1	0
1	$1 \rightarrow 0$	1	0	1	0	0
1	$0 \rightarrow 1$	1	0	0	0	1

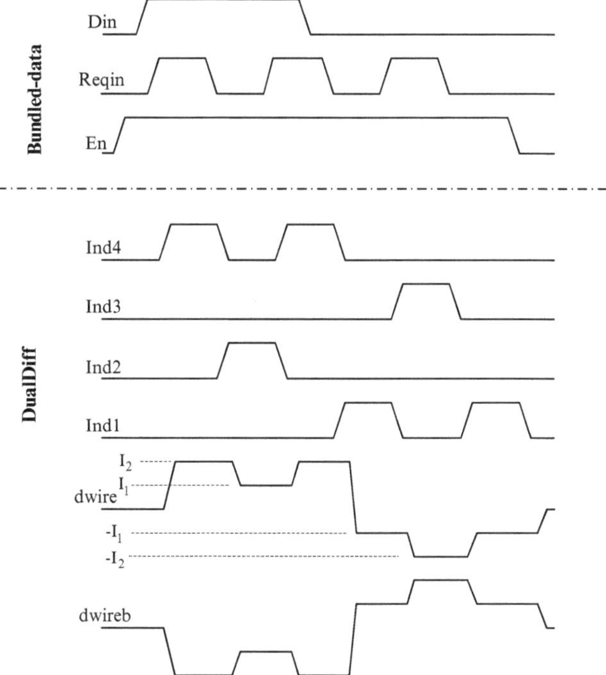

Fig. 4.27 Communication protocol of *Dualdiff* interconnect

signaling features. A change in the current level on the wire indicates arrival of new data (delay-insensitivity), while the direction of the current flow reveals the logical value of the transmitted bit. The encoding protocol is shown in Table 4.5. When the transmitted bit is 1 the driver sources current to one of the wires and sinks current from the other wire, and vice versa for bit 0 transmission. The value of the current on the wires switches between I_1 and I_2 at every new transmission event. The communication protocol of this interconnect is shown in Fig. 4.27.

The encoder receives as inputs data (*Din*) and request (*Reqin*) signals in the bundled data form. It encodes the data and request together and outputs voltage pulses which serve as inputs to the driver. The encoder circuit is shown in Fig. 4.28. The *En* signal is used for enabling transmission and it is an active-high signal. That is, it is *high* during transmission and *low* during idle periods. The edge sensitive flip-flops sample the data input at the edges of the bundled-data two-phase request

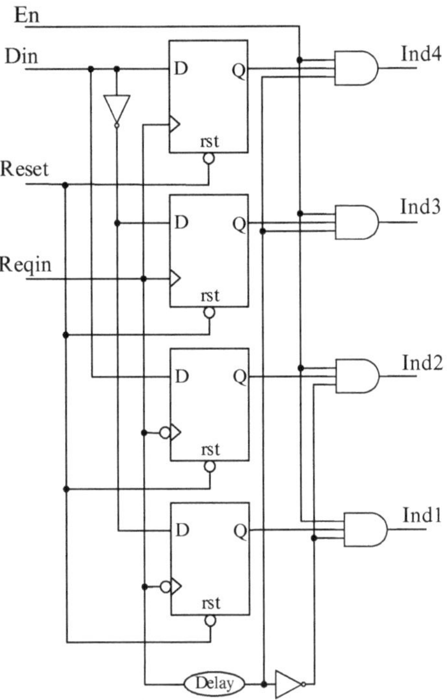

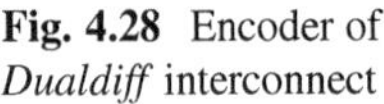
Fig. 4.28 Encoder of *Dualdiff* interconnect

input (*Reqin*). Only one of the AND gates (*N1* to *N4*) outputs voltage pulse at every transmission event. The encoder outputs, *Ind1*, *Ind2*, *Ind3* and *Ind4* are inputs to the differential driver.

4.3.2 Driver, Receiver and Completion Detector

The designs of driver, receiver and completion detector circuits are presented in this section and shown in Figs. 4.29 and 4.30. As shown in Fig. 4.29, source coupled CMOS bipolar current-mode drivers are used. Such a driver conveys two currents of the same amplitude but opposite polarity to the wires such that not only the effect of supply voltage fluctuations on the wires is minimized, but also the noise injection from the driver to the substrate is minimal. Two bipolar current-mode drivers are used in order to drive the two current levels I_1 and I_2. At every transmission event, the driver sources current to one of the wires and sinks the same amount of current from the other wire.

The receiver senses the direction of the current flow to retrieve the transmitted data. The receiver of this interconnect has two stages, a current direction sensor and a differential amplifier as shown in Fig. 4.30. The current direction sensing circuit is a modified version of the one presented in [54]. In our design, the termination transistor is diode connected to ensure its saturation operation and also to mirror the

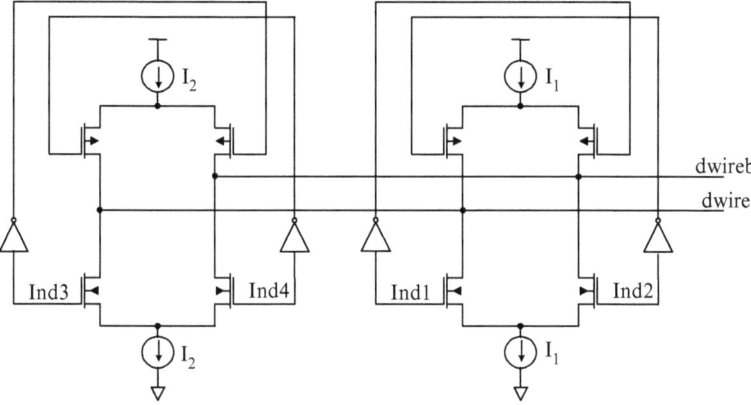

Fig. 4.29 Driver of *Dualdiff* interconnect

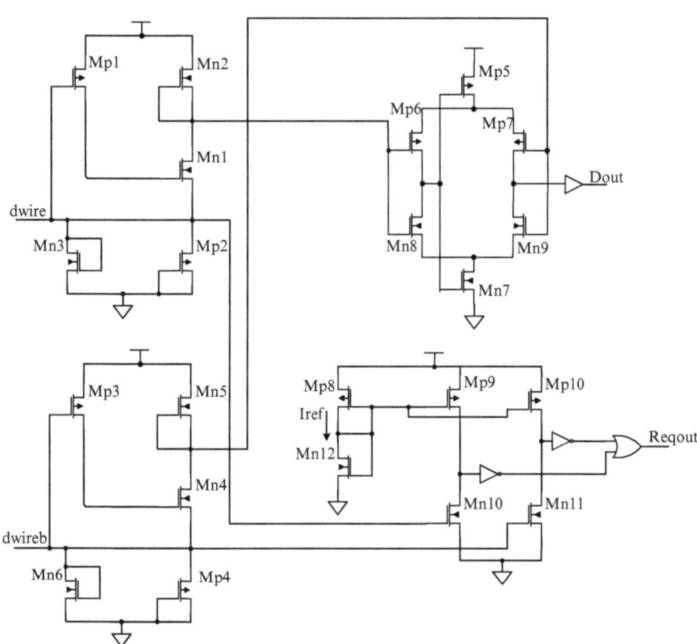

Fig. 4.30 Receiver and completion detector of *Dualdiff* interconnect

wire current. Furthermore, unlike in [54] the current sensor output is not connected back to the termination transistor gate to avoid its effect on output switching. Thus, the *Mn3* and *Mn6* transistors provide a low impedance path to ground for current sourced by the driver and serve as best matched impedance termination since they operate in saturation. Consider the top current sensor in Fig. 4.30. The transistor *Mp1* provides negative feedback to transistor *Mn1*. It turns the gate of *Mn1* on and off as required and helps in modulating the input impedance. The transistor *Mp2* provides a constant current bias hence regulating the transconductance of *Mn1*. The source terminal of the transistor *Mn1* is connected to *dwire*. When current is sunk by the driver *Mn1* becomes *on* and pulls the output of the current sensor to *low*. When current is sourced by the driver, the source voltage of *Mn1* rises thus turning it *off*. In this case the current flows through the load transistor *Mn2* to the output making the output voltage of the current sensor *high*.

The output of the current sensor is not full swing. In addition, the receiver needs to have high common-mode noise rejection capability in order to take full advantage of differential signaling. Due to these, the second stage, a high-speed self-biased differential amplifier is used. The amplifier consists of source coupled NMOS and PMOS transistors (*Mn8*, *Mn9*, *Mp6*, and *Mp7*). It operates at high speed because its output switching currents are significantly greater than its quiescent current. It has also a higher differential-mode gain compared to conventional amplifiers and a large common-mode input range because its bias condition adjusts itself to accommodate the input swing [91, 95]. The bundled-data input to the receiving module, *Dout*, is the output of the amplifier without requiring additional data decoding logic.

Since the receiving side has a bundled-data interface, it requires a separate data validity indicator (Reqout in Fig. 4.30). Based on the encoding, it is known that the current on the wire becomes $\pm I_2$ when the request from the sending module is *high* and $\pm I_1$ when the request is *low*. To decode the request signal the output currents of both wires are compared separately with the reference current using a current comparator as shown in Fig. 4.30. The currents in *dwire* and *dwireb* are mirrored into transistors *Mn10* and *Mn11*, respectively. These mirrored wire currents will be compared with the reference current, which is generated by diode connected *Mp8* and *Mn12* transistors. The reference current value is $0.5 * (I_1 + I_2)$. If either of the wire output currents is greater than the reference current, *Reqout* will be *high* and if the current in both wires is less than I_{ref} then *Reqout* will be *low*.

4.3.3 Acknowledgment Transmission

In this interconnect design, an acknowledgment is sent for each transmitted bit from the receiving module. The voltage-mode bundled-data acknowledgment signal (*Ackin*) is converted into a differential current signal during transmission and back into a voltage-mode signal (*Ackout*) at the transmitter side. The acknowledgment transmission also uses differential current sensing signaling. The driver and receiver circuits of this transmission along with distributed RLC model of the acknowledgment wire is shown in Fig. 4.31. The current through the differential acknowledgment wires, *Ackwire* and *Ackwireb* becomes $+I$ and $-I$, respectively when *Ackin* is *high* and vice versa when *Ackin* is *low*. The two current direction sensors, which are the same as in data receiver, detect the direction of current flow. The self-biased differential amplifier retrieves the transmitted acknowledge signal using the output of the current sensors.

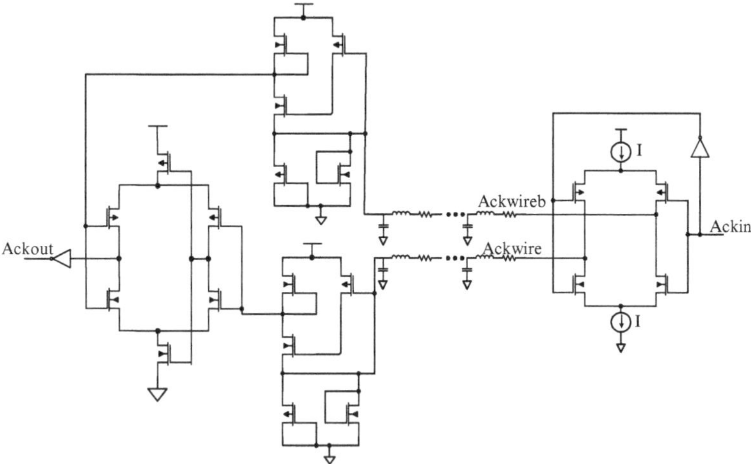

Fig. 4.31 Acknowledgment transmission of *Dualdiff* interconnect

4.3.4 Simulation Results and Analysis

The wire properties were set according to the ITRS 65 nm technology node for global wiring. RLC matrices of the wire were extracted using FastHenry and Linpar fieldsolvers. During extraction, both wire width and separation distance were set to 210 nm and the wire thickness was set to 242 nm. In the interconnect simulation, the wires were modeled as a distributed RLC using the extracted per unit values. The circuits were designed and simulated in Cadence Analog Spectre using 65 nm CMOS technology from STMicroelectronics and 1 V supply voltage. The simulation waveforms are shown in Fig. 4.32. It consists of the input data and request, current and voltage of the wire, the amplifier output, the request output, and acknowledgment input and output.

A reference interconnect is designed and simulated to compare the performance, power consumption and area of the *Dualdiff* interconnect. In order to determine the contribution of the novel integration scheme in *Dualdiff*, the reference interconnect uses conventional integration of delay-insensitive data transfer with differential current sensing signaling. It uses LEDR data encoding due to its simpler data decoding and completion detection schemes. It requires four wires per bit, that is, two for differential phase and the other two for differential state transmissions as in [83]. The same signaling circuits are used as in the *Dualdiff* interconnect to have proper comparison. The reference LEDR encoded differential current sensing interconnect (*LEDRdiff*) is shown in Fig. 4.33. It requires two bipolar differential current-mode drivers, four current direction sensors, and two differential amplifiers. Its data encoding and data validity indicator decoding circuits are also shown in Fig. 4.33.

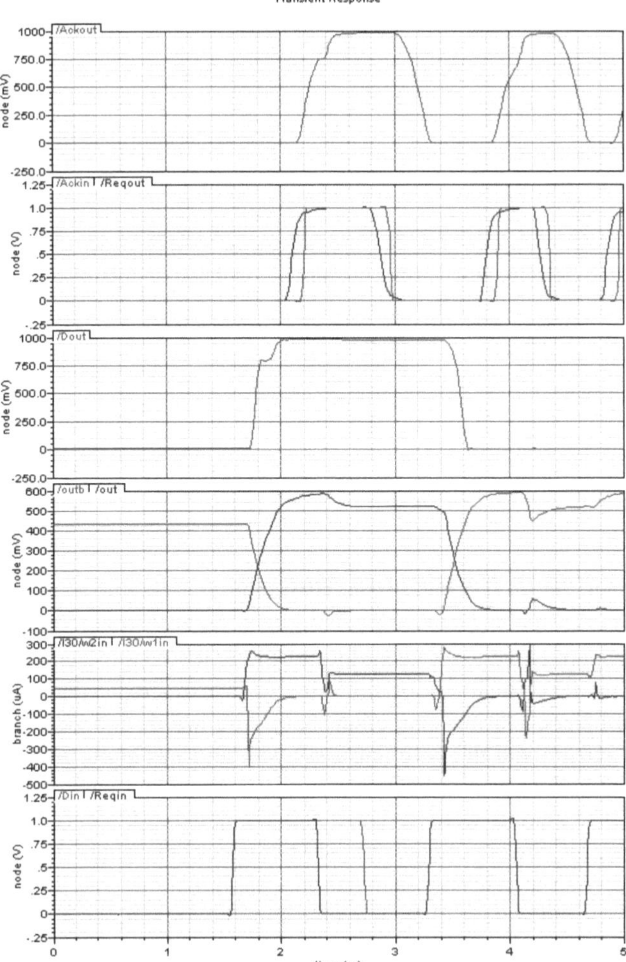

Fig. 4.32 Simulation waveforms of *Dualdiff*

The forward latencies of both interconnects when the wire length varies from 1 to 5 mm are shown in Fig. 4.34. The latency of the proposed interconnect is much smaller than that of the reference one. For example, at 2 mm wire length its latency was less than one-half of that of the reference. As shown in Fig. 4.35, the throughput of the *Dualdiff* interconnect is 1.92 and 1.54 times that of the reference interconnect for 1 and 5 mm long links, respectively. It has a throughput of 1.34 Gbps at 5 mm wire length. The performance penalty of the reference interconnect comes mainly from its encoding and completion detection circuits.

The power consumption of the proposed interconnect is smaller than that of the reference interconnect as expected (Fig. 4.36). At 5 mm wire length 24% power

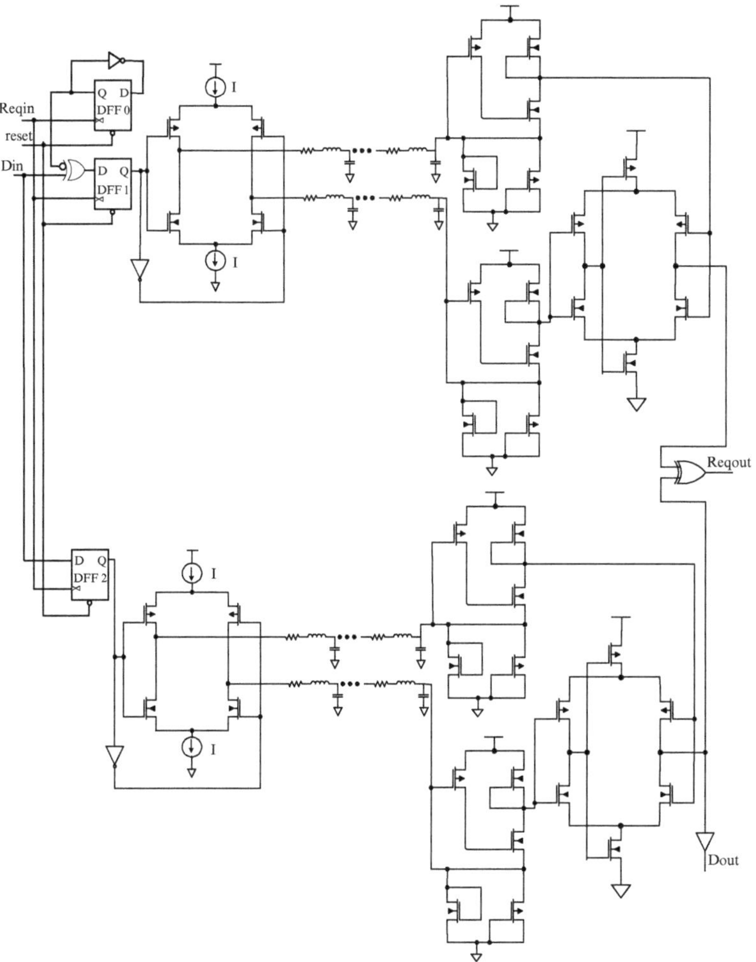

Fig. 4.33 LEDR encoded differential interconnect

savings has been gained compared to the reference interconnect. The average energy dissipated per every transmitted bit is also examined for both interconnects and is shown in Fig. 4.37. The energy per bit of *Dualdiff* interconnect is much smaller than the reference energy dissipation. Its energy per bit is less than one-third and one-half of that of the reference at 1 and 5 mm wire lengths, respectively. The reference interconnect energy dissipation increased much faster than the proposed one for longer wires. All these analyses show the superiority of the proposed interconnect over the conventional delay-insensitive differential interconnect. Moreover, the proposed interconnect is area efficient since it reduces the required number of wires by half. It requires 15% less active area and 40% less wiring area for 2 mm long interconnect (see Table 4.6).

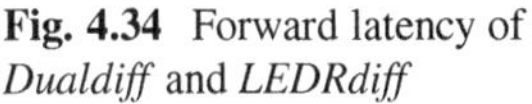
Fig. 4.34 Forward latency of *Dualdiff* and *LEDRdiff*

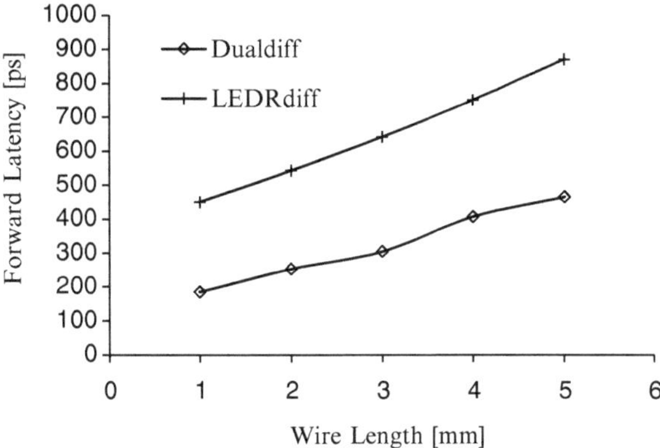

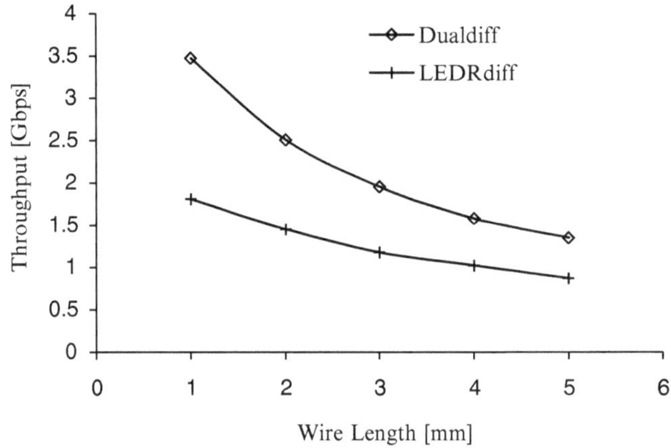

Fig. 4.35 Throughput of *Dualdiff* and *LEDRdiff*

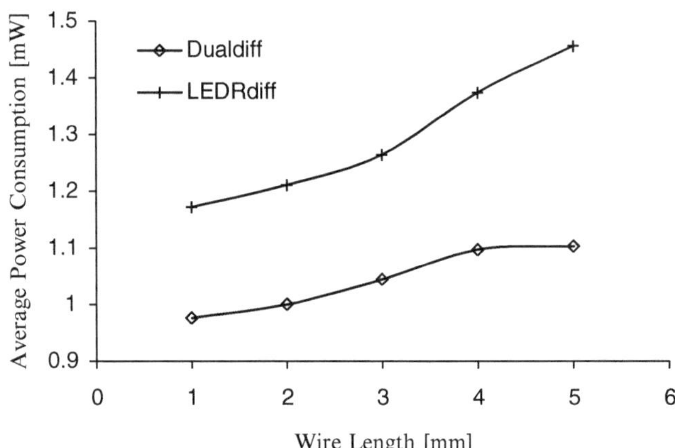

Fig. 4.36 Power consumption of *Dualdiff* and *LEDRdiff*

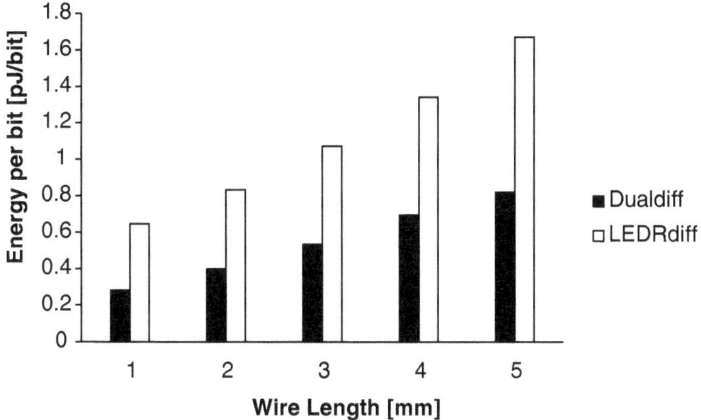

Fig. 4.37 Energy per bit dissipation of *Dualdiff* and *LEDRdiff*

Table 4.6 Area comparison between *Dualdiff* and *LEDRdiff*

Interconnect	Active area [μm^2]	Wiring area [μm^2]
Dualdiff	16	2,520
LEDRdiff	19	4,200

4.4 Chapter Summary

The design and analysis of three high-performance and delay-insensitive global on-chip interconnects were presented. The delay-insensitivity makes the communication robust and attains average-case performance rather than worst-case which is the situation in communication based on timing constraints. The first interconnect presented in this chapter, *LEDRCm*, has achieved higher throughput and dissipated lower energy per bit than the conventionally implemented *LEDRVm* and *TPDRVm* interconnects. The second interconnect presented in this chapter, *PMCm*, uses two-phase 1-of-4 encoding and multilevel current sensing signaling. The performance analysis showed that *PMCm* has achieved higher throughput and lower latency than its two reference interconnects, *TPVmP* and *TPVmRep*. In addition, the energy per bit dissipation of *PMCm* was lower than that of the *TPVmRep*. The last interconnect presented in this chapter is based on a novel integration of delay-insensitive encoding and differential current sensing signaling. Only half number of wires are required compared to conventional integration of the two schemes, making it both area and power efficient. It has achieved higher performance than a reference interconnect, *LEDRdiff*. It has also consumed lower power and dissipated lower energy per bit than the *LEDRdiff*. Therefore, the presented three on-chip interconnects are prominent candidates for high-performance, energy efficient and delay variations tolerant global on-chip communication.

Chapter 5
Enhancing Completion Detection Performance

In the previous chapter, designs of high-performance and delay-insensitive current sensing interconnects have been presented. In delay-insensitive transmission, validity of the data is encoded within the data itself at the transmitter, and the data validity test, i.e., completion detection, as well as data decoding is performed at the receiver. The delay incurred due to completion detection increases with bit width of the transmission channel and affects the performance of the communication significantly. In order to overcome this overhead, a high speed completion detection technique along with its CMOS implementation is designed and presented in this chapter. Unlike the conventional detection circuits, the delay of the presented completion detection circuit is not affected by the bit width of the channel. This optimizes the performance of delay-insensitive current sensing links further since it was already demonstrated in Chap. 4 that the current sensing interconnects achieve higher performance and better power efficiency than the interconnects using voltage-mode signaling with repeaters or pipelines.

The chapter is organized as follows. First delay-insensitive bit parallel transmission and the overhead of completion detection in such transmission are discussed. The novel high speed completion detection technique and its implementation details are presented in Sect. 5.2. Two delay-insensitive links, which use the proposed completion detection technique are presented as case studies in Sects. 5.3.1 and 5.3.2. The design of the acknowledgment interconnect for the case study links are discussed in Sect. 5.3.3 and the design of reference links are explained briefly in Sect. 5.4. In Sect. 5.5 simulation details and analysis of performance, power, energy, noise effects and area of the case studies as well as reference links are presented. The summary presented in Sect. 5.6 concludes the chapter.

5.1 Delay-Insensitive Bit Parallel Transmission

A GALS communication method is used in almost all proposed NoC designs and is expected to be an attractive approach to overcome many of the timing problems [55]. The GALS approach simplifies clock tree design and results in easily scalable

E.E. Nigussie, *Variation Tolerant On-Chip Interconnects*, Analog Circuits and Signal Processing, DOI 10.1007/978-1-4614-0131-5_5,

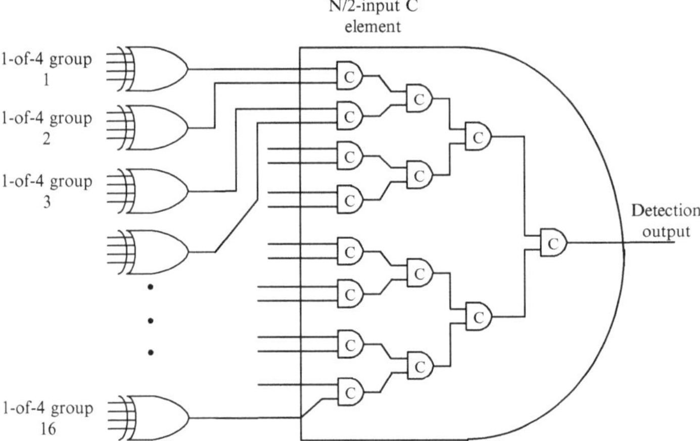

Fig. 5.1 Completion detector of 32-bit two-phase 1-of-4 transmission

clocking systems. It also enables better energy savings since each functional unit can easily have its own independent clock and voltage [56]. Furthermore, it allows easy implementation of a distributed power management system for the entire chip [57]. A fully self-timed NoC in the GALS clocking scheme gives a better network saturation threshold, smaller average power consumption, slightly higher maximal bandwidth and much smaller packet latency (2.5 times smaller) than the multi-synchronous NoC implementation [58]. A number of fully asynchronous GALS NoCs have been proposed and implemented, such as MANGO [86], ANoC [87], ALPIN [81], FAUST chip [88] and QNoC [89]. Hence, due to the advantages of a fully self-timed GALS NoC and given its ability to work reliably in the presence of variations, a self-timed delay-insensitive link between NoC routers is a natural choice.

In bit parallel transmission, the throughput of a delay-insensitive link decreases when the bit width of a channel increases, because of the increase in the delay of completion detection. Conventionally, completion detection is carried out by sensing either voltage transitions or levels on each data wire. This requires logic circuitry whose delay increases drastically when the channel bit width increases, causing a bottleneck to achieve high performance communication using a delay-insensitive interconnect. The completion detection logic for two-phase dual-rail and 1-of-4 encoded links are shown in Figs. 5.1 and 5.2 for 32-bit parallel transmission. These two encodings are the simplest and the most commonly used on-chip delay-insensitive codes [59], requiring two signal wires per each transmitted bit. In 32-bit 1-of-4 encoded transmission, the delay incurred due to the completion detection is the sum of delays of a 4-inputs XOR gate and four 2-input C-elements, see Fig. 5.1. The completion detection logic of 32-bit two-phase dual-rail encoded transmission has an extra delay of a 2-input C-element compared to 1-of-4 as shown in Fig. 5.2. On the other hand, its XOR gates have only 2-inputs. In 65 nm technology the

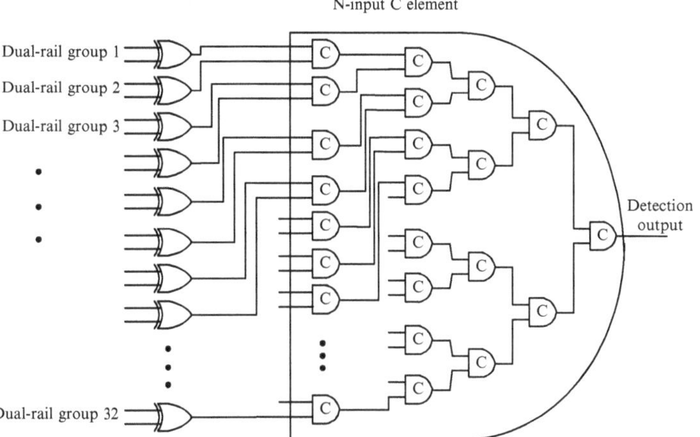

Fig. 5.2 Completion detector of 32-bit two-phase dual-rail transmission

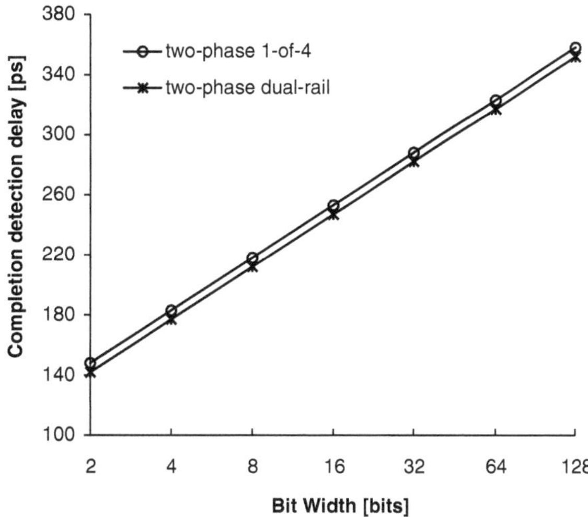

Fig. 5.3 Completion detection delays of interconnects in 65 nm technology

completion detection delay of 32-bit two-phase dual-rail encoded transmission is 217 ps. For example, if the data receiving block runs at 5 GHz, it has to wait more than one clock cycle only because of the completion detection. The delay of completion detection for N-bit two-phase 1-of-4 encoded transmission is the sum of a 4-input XOR gate delay and $(log_2 N - 1)$*2-input C-element delay. In case of two-phase dual-rail encoded N-bit transmission, the delay of completion detection is the sum of a 2-input XOR gate delay and $log_2 N$*2-input C-element delay. So, the larger the channel bit width is, the longer is the overall time spent in completion detection, because each detection circuit becomes a tree of logic elements. The increase in this delay penalty for two-phase dual-rail and 1-of-4 encoded channels designed in 65 nm technology is shown in Fig. 5.3.

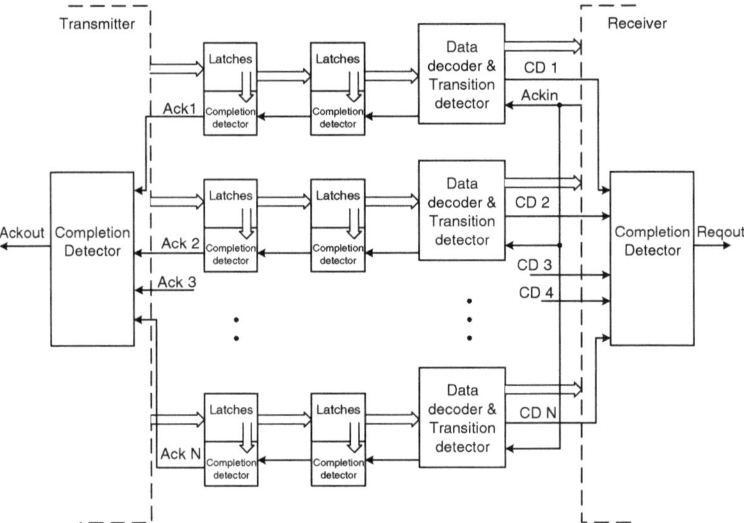

Fig. 5.4 Completion detection in a pipelined voltage-mode link

Traditionally, optimal repeater insertion together with pipelining is the method to achieve high throughput in global voltage-mode on-chip interconnects. If such a pipelined interconnect is delay-insensitive, each pipeline stage including the transmitter and receiver themselves, requires area, power, and time consuming completion detection logic. At the receiver side completion detection is needed to indicate the validity of the arrived data and at the transmitter side to indicate the acceptance of the transmitted data, since in a pipelined channel an acknowledgment is sent per group instead of for the whole channel. For example, in [61] an acknowledgment is sent per each 1-of-4 group, this helps to reduce the speed penalty due to large detection logic at each pipeline stage. A block level diagram of a delay-insensitive pipelined voltage-mode link is illustrated in Fig. 5.4 showing delay causing completion detection blocks at the pipelines latches, receiver and transmitter.

A delay-insensitive current sensing link does not require repeaters nor pipelining to boost its throughput, indicating that completion detection is carried out only once at the receiver, and therefore it achieves higher performance and better power efficiency than a pipelined voltage mode interconnect. This has been proved in Chap. 4 where, however, wire currents are first converted to voltages, and the actual completion detection is carried out in the voltage mode, resulting in a significant speed penalty. Hence, a high speed completion detection technique which uses the wire currents directly without conversion to voltage mode and carries out completion detection in current mode is proposed and presented in this chapter. Unlike with the conventional completion detection logic, delay of the proposed scheme does not increase with the link bit width, it is bit width independent.

5.2 High-Speed Completion Detection Technique

The proposed completion detection technique uses directly the current on each data wire and carries out completion detection in the current mode. The idea is to sum the currents on all the data wires of a channel and then compare this sum current to a reference current. Implementation of this technique requires only current mirrors, a current source, and a current comparator. The comparator takes as inputs the sum current and the reference, and outputs a full-swing completion detection signal. This signal becomes *high* when the sum current is greater than the reference current, indicating the validity of every received data signal. Unlike with the conventional voltage mode scheme, the speed of the proposed scheme is not affected by the channel bit width, because the current summation is carried out by wiring and its delay is only due to comparing currents.

The completion detection circuit is shown in Fig. 5.5. It supports detection of dual-rail (1-of-2) and 1-of-4 encoded data, both of which use *2N* wires to convey *N*-bit data, so that the number of active wires per transmission is *N* in the 1-of-2 case and $N/2$ in the 1-of-4 case. The diode connected transistors *Mw(1)* to *Mw(2N)* (one transistor per wire) are used to input the currents on the wires and mirror them to *Ms(1)* to *Ms(2N)* transistors, respectively, which are connected together to generate the sum current $I(sum) \equiv Code \times (S^{-1}) \times N \times I(w)$. Here *I(w)* is the nominal current on a single data wire, *N* is the number of bits, *S* is the current down-scaling factor ($S \geq 1$) indicating the current drive ratio between the transistors *Mw(i)* and *Ms(i)*, and *Code* is either 1 (1-of-2 code) or 0.5 (1-of-4 code) indicating the number of active wires per transmitted bit. By using a scaling factor (*S*) larger than 1 the power consumption of the circuit can be efficiently reduced. The reference current,

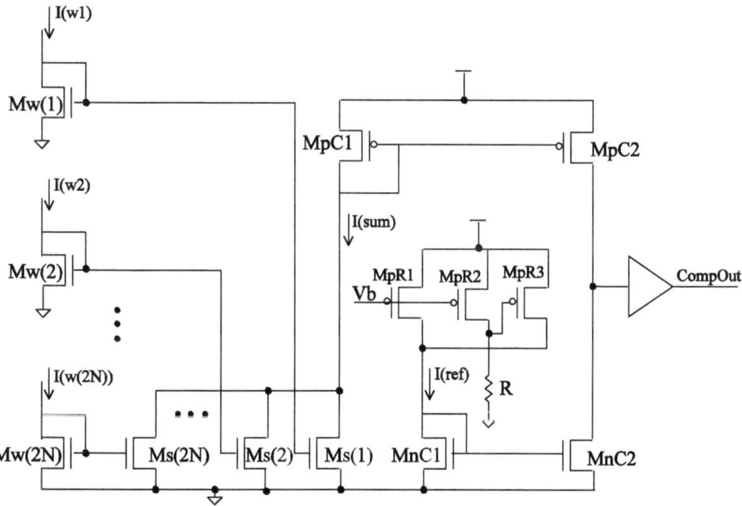

Fig. 5.5 High-speed completion detection circuit

$I(ref)$, is generated using an addition based process invariant current source [62]. Its value is $I(ref) \equiv Code \times (S^{-1}) \times (N - 0.25) \times I(w)$. This value is chosen in order to compare after the wire current of the last bit reaches 75% of its current. The comparator transistor *MpC1* mirrors $I(sum)$ to $MpC2$, and *MnC1* mirrors $I(ref)$ to $MnC2$. The comparator output becomes *high* (*low* otherwise), when the current of $MpC2$ is greater than that of $MnC2$. Due to process and supply voltage variations, the sum and reference currents may vary from their nominal values affecting the reliability of completion detection. In order to have correct operation, condition 5.1 has to be fulfilled.

$$\Delta(I(sum)) + \Delta(I(ref)) < (S^{-1}) \times \left(\frac{I(w)}{2}\right) \tag{5.1}$$

$$\Delta(I(sum)) \cong Code \times (S^{-1})\, N \times I(we) \tag{5.2}$$

For *I(sum)* the variation can be expressed as in Equation 5.2, where *I(we)* is the worst case variation of the current on a single wire. For *I(ref)*, according to an extensive Monte Carlo analysis of the circuit for $N = 2$ to 64-bit, $I(w) = 200\,\mu\text{A}$ to $300\,\mu\text{A}$, and $S = 1$ to 5, it is safe to assume that the variation is within the bound expressed in Eq. 5.3.

$$\Delta(I(ref)) < (S^{-1}) \times \frac{I(w)}{6} \tag{5.3}$$

Substituting Eqs. 5.2 and 5.3 into Eq. 5.1 and solving for N yields the following constraints:

$$N < \frac{1}{3} \times SNR \tag{5.4}$$

$$N < \frac{2}{3} \times SNR \tag{5.5}$$

where *SNR* is the signal-to-noise ratio of a single data wire and expressed in Equation 5.6. Condition 5.4 is for the 1-of-2 (Code = 1) and Condition 5.5 is for 1-of-4 (Code = 0.5) codes.

$$SNR = \frac{I(w)}{I(we)} \tag{5.6}$$

The higher the *SNR* is the larger number of bits (N) can be reliably transmitted and detected. Furthermore, for a given *SNR*, a 1-of-4 encoded channel can be twice as wide as a 1-of-2 encoded channel, because the number of active wires is half of that of the 1-of-2 case. The relation between N and *SNR* for the 1-of-2 and 1-of-4 encoded channels is shown in Fig. 5.6.

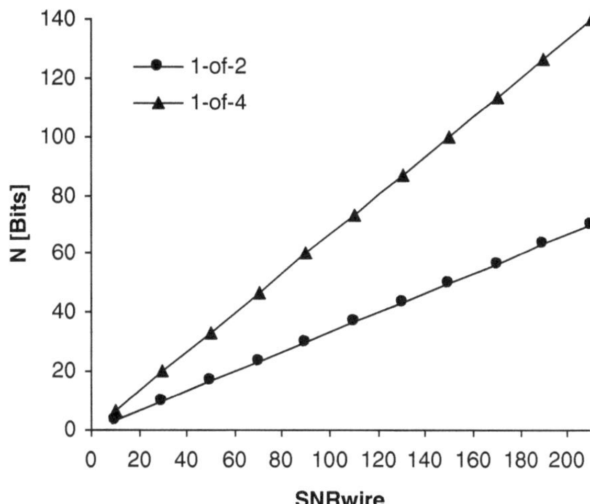

Fig. 5.6 Bit-width versus SNR of a wire for a reliable detection

5.3 Case Studies

In this section, the redesign of *PMCm* (Sect. 4.2) and *Dualdiff* (Sect. 4.3) interconnects in order to use the proposed completion detection technique is presented. In the initial design of these two interconnects, the wire currents have been converted to voltages and then conventional completion detection has been carried out in voltage mode. In the analysis section, the performance improvement due to the novel completion detection will be presented.

5.3.1 1-of-4 Encoded Current Sensing Interconnect

As already discussed in Sect. 4.2, in the *PMCm* interconnect the two-phase bundled-data voltage-mode data and request signals are converted into pulsed 1-of-4 multilevel current sensing signaling at the transmitter side. At the receiver side, delay-insensitive current sensing signaling is turned back into bundled-data voltage-mode communication. On the wires information is represented as current rather than voltage transitions, one of the four data wires draws current to indicate the presence of a new two-bit data symbol. The current detected at the receiver has three different values: 0, I_1, and I_2. The values I_1 and I_2 are used when the voltage-mode request signal *Reqin* at the transmitter side is *low* and *high*, respectively, reflecting the adopted two-phase communication protocol. The value 0, in turn, means that there is no symbol on a wire, representing the idle period. Since the design of encoder, driver, receiver and data decoder of this interconnect is the same as in Sect. 4.2, only the completion detection implementation is discussed here. The *PMCm* interconnect which uses the proposed completion detection technique is called *PMCmFCD*.

The completion detection circuit for a 4-bit transmission using *PMCmFCD* interconnect is shown in Fig. 5.7. This detector requires two current comparators because of the power saving scheme of the *PMCm* interconnect. That is, during the idle period of the transmission the currents on the wires are switched off. This switching off should not affect the state of the two-phase bundled-data *Reqout* signal, which is the output of the completion detector. The main comparator, composed of *Mp2* and *Mn2*, compares the sum of wire currents with a reference current. For N-bit transmission, this reference current I_{ref} is in the range of $S \times N/2 \times I_1 < I_{ref} < S \times N/2 \times I_2$, where S is the wire current scaling factor. If the current through the transistor *Mp2* is greater than the reference current in the transistor *Mn2*, the output *C1* goes *high*, otherwise it goes *low*. *C1* is latched to the output of the completion detector only when there is current on the wires. To determine the availability of current on the wires, an additional comparator is required. This comparator compares the sum of wire currents with a small reference current I_{ref1}, which is in the range of $0 < I_{ref1} < S \times N/2 \times 0.5 \times I_1$. As long as the current in the transistor *Mp3* is greater than I_{ref1} (the reference current that is mirrored to transistor *Mn3*), the output *C1* is latched to *Reqout*. If the current in *Mp3* is less than I_{ref1}, the output *C0* becomes low, which in turn causes the latch to enter the hold mode, and there is no change in the *Reqout* signal. *Winv* in Fig. 5.7 is a weak inverter which is used as a keeper.

The major performance improvement due to the use of this completion detection scheme becomes significant when the number of bits transmitted in parallel increases. In *PMCm*, where the completion detection is carried out in the voltage mode after converting the wire current, one 1-of-4 group (2-bit transmission) detector requires eight current comparators, two 4-input NAND gates, a 2-input NAND gate and a 2-input C-element. For N-bit transmission, it requires $(N/2)$ times the components of one 1-of-4 group detector in addition to the $(N/2)$-input C-element. Thus, the speed penalty due to the completion detection becomes considerable especially in a high performance system. For example, in 65 nm technology, the delay of a 64-bit transmission detector of *PMCm* is 124 ps, whereas in *PMCmFCD* it is only 52 ps. Detailed analysis will be presented in Sect. 5.5.

5.3.2 Dual-rail Encoded Differential Current Sensing Interconnect

In this interconnect, differential current sensing signaling and two-phase dual-rail encoding are integrated in an area and power efficient manner. Both current directions and current values are used simultaneously to get both delay-insensitivity and differential signaling features. A change in the current level on the wire indicates arrival of new data (delay-insensitivity), while the direction of the current flow reveals the logical value of the transmitted bit. As already discussed, retrieving the data validity signal is necessary because the receiving module has a bundled-data encoded interface. The arrival of new data on the wire is indicated by either of

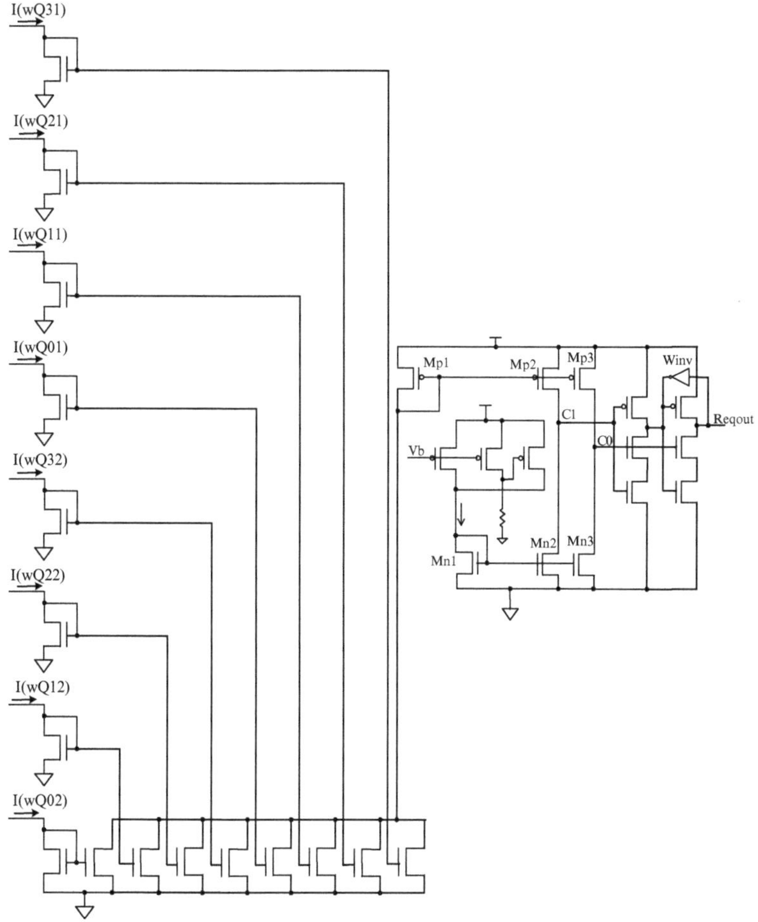

Fig. 5.7 Completion detection circuit of 4-bit *PMCmFCD* link

the two wire current levels I_1 and I_2, enabling the use of the proposed completion detection technique as a more appropriate choice. This is because it is fast, consumes less power and area compared to the *Dualdiff* interconnect (Sect. 4.3) where each wire current is first sensed, then converted to voltage and detection is carried out in the voltage mode. The *Dualdiff* interconnect which uses the proposed completion detection technique is here called *DualdiffFCD*.

In *DualdiffFCD* the current on the wire becomes I_2 when the request input from the sending module is *high* and it becomes I_1 when the request input is *low*. During the idle period of transmission, the wire current is switched off to save power. The completion detection circuit of *DualdiffFCD* is shown in Fig. 5.8 for 4-bit transmission. It is the same as the completion detection circuit of *PMCmFCD* since both use three current levels, 0, I_1 and I_2.

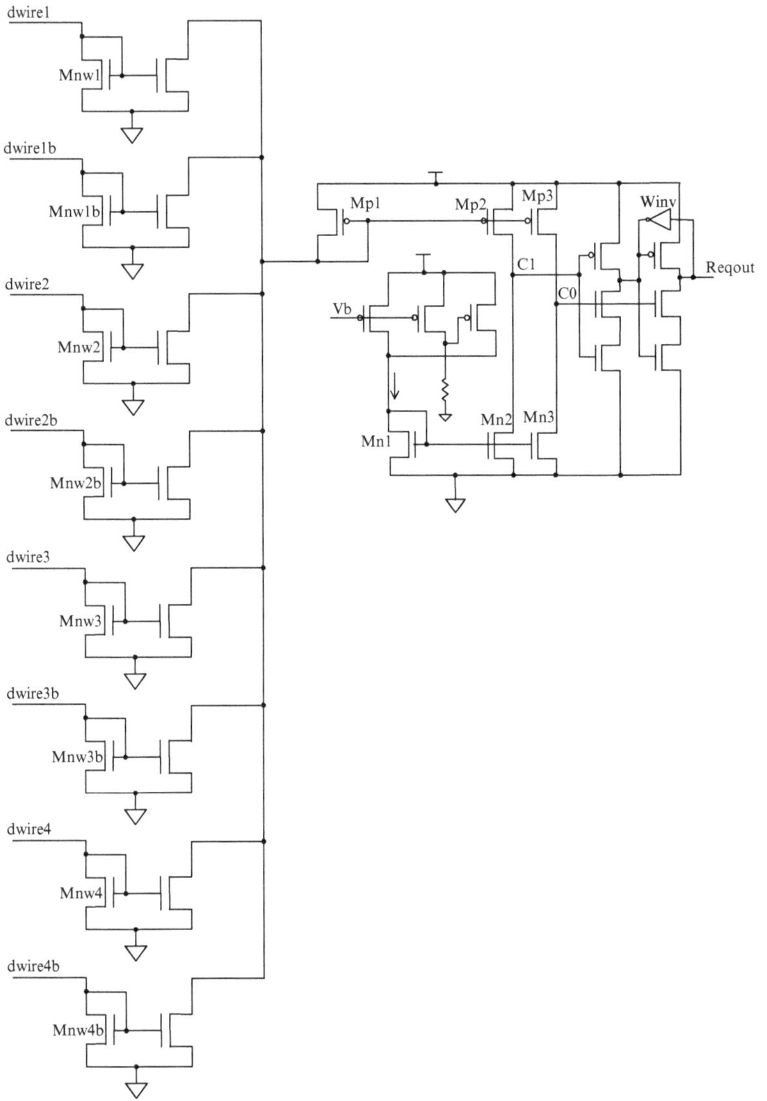

Fig. 5.8 Completion detection circuit of 4-bit *DualdiffFCD* link

5.3.3 Acknowledgment Transmission

The designs of acknowledgment transmission circuits are the same for the two case study interconnects. The voltage mode bundled-data acknowledgment signal (*Ackin*), sent by the receiver module, is converted into a current mode signal during transmission and back into a voltage mode signal (*Ackout*) at the transmitter side. Current sensing differential signaling is used.

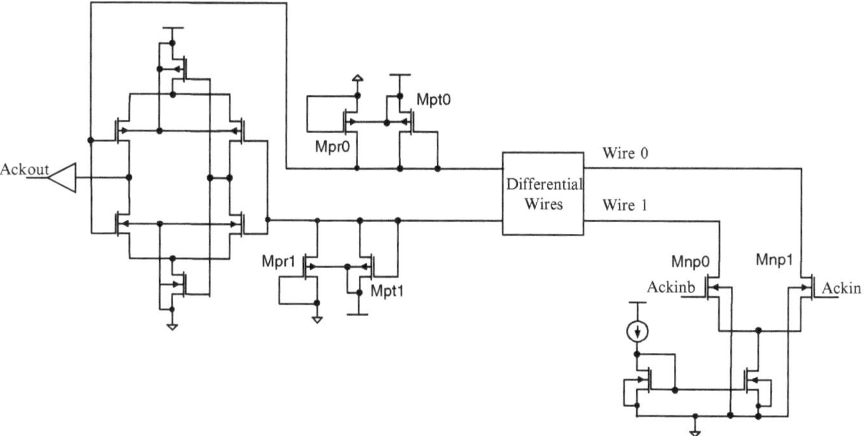

Fig. 5.9 Acknowledgment signal transmission

A source-coupled differential current-steering driver, shown in Fig. 5.9, is used. It is fast because it has an extremely sharp transient response. It has also an advantage of reducing the AC component of the power supply noise because the circuit draws constant current from the supply. The complementary outputs of the driver are attached to the differential pair of wires. The other end of the transmission is parallel terminated into a positive voltage. When *Ackin* makes a transition from *low* to *high*, there is current in *wire0* and no current in *wire1*. When *Ackin* makes a transition from *high* to *low* there is current in *wire1* and no current in *wire0*. Diode connected *Mpt0* and *Mpt1* transistors are used as termination loads. The transconductance of these transistors is regulated through the use of *Mpr0* and *Mpr1*. The receiver is a high speed self-biased differential amplifier, which has a high differential-mode gain.

5.4 Reference Cases

Four interconnects are designed to serve as reference cases. Two of them are used to determine the performance improvement enabled by the presented high speed completion detection technique. They are the ones presented in Sects. 4.2 and 4.3, where the wire currents are converted into voltages and completion detection is carried out in the voltage mode. The purpose of the other two is to analyze the contributions of the current sensing signaling along with the high speed completion detection. They use optimally pipelined voltage-mode signaling. The pipeline stages are inserted at distances where optimal throughput can be achieved. One uses two-phase 1-of-4 encoding (*1of4VmP*) and the other uses two-phase dual-rail encoding (*DualVmP*). The implementation details of the *1of4VmP* interconnect are mostly the same as in Sect. 4.2.5. The only difference is that here the pipeline stages are inserted at optimal distances, where the highest possible throughput can be achieved. The encoding, decoding and completion detection of the *DualVmP* circuits are the same as the ones presented in Sect. 4.1. Their gate-level implementations were shown in Figs. 4.2 and 4.3. In the two pipelined links, one acknowledgment wire per 2-bit transmission is used to minimize the delay due to completion detection. With this configuration, completion detection is carried out per each 2-bit group at each pipeline stage and for all bits only at the receiver and transmitter sides.

5.5 Simulation Results and Analysis

5.5.1 Wire Model

The wire properties are set according to ITRS 65 nm technology node for global wiring [63]. The RLC values of the wires are extracted using field solvers from microstrip configuration. In the model, both wire width and separation distance are set to 210 nm (two times minimum pitch for global wiring) and the wire thickness is set to 242 nm. The resistance and inductance matrices are extracted using FastHenry [45], while the capacitance matrices are extracted using Linpar [46].

A number of data transmission wires with different bit widths and with both capacitive and inductive coupling are modeled to analyze the performance enhancement due to the proposed completion detection scheme. Parallel data transmission of 2, 4, 8, 16, 32 and 64-bits are considered. To implement these transmissions in the current sensing interconnects 2, 4, 8, 16, 32, 64 and 128 parallel wires with coupling are modeled and used in the simulation. Their acknowledgment transmission wire is assumed to be shielded at both sides with grounded wires and hence there is no coupling between the data and acknowledgment wires. In case of the two pipelined reference links 5, 10, 20, 40, 80, and 144 parallel wires with both capacitive and inductive coupling are modeled and used in the simulations. The required number of wires is different for the pipelined links since an acknowledgment is sent for each 2-bit group. Performance, power consumption and energy dissipation per bit of the interconnects are examined by varying the wire length from 1 to 5 mm during simulations.

5.5.2 Simulations Setup

All simulations are carried out in Cadence Analog Spectre using 65 nm CMOS technology from STMicroelectronics and 1V supply voltage. A number of simulations are carried out in order to analyze and compare the performance and power consumption of the presented on-chip interconnects including the references. In order to avoid confusion between the interconnects the following naming conventions are used. The current sensing two-phase 1-of-4 encoded and two-phase dual-rail encoded differential links with the proposed completion detection are called *PMCmFCD* and *DualdiffFCD*, respectively. The current sensing two-phase 1-of-4 encoded and two-phase differential dual-rail encoded interconnects, which use the conventional completion detection technique are called *PMCm* and *Dualdiff*, respectively. The voltage-mode pipelined two-phase 1-of-4 and dual-rail encoded interconnects are called *1of4VmP* and *DualVmP*, respectively.

The *PMCmFCD* and *DualdiffFCD* interconnects are simulated for 2-bit transmission by varying the wire length from 1 to 5 mm. They are also simulated for 2-, 4-, 8-, 16-, 32- and 64-bit transmission with the wire length of 2 mm. The 2 mm distance is chosen because there is a manufactured prototype in 65 nm technology from Intel where the distance between neighboring tiles is 2 mm [69]. Furthermore, *PMCm* and *Dualdiff* interconnects are redesigned and simulated in 65 nm technology for 2-, 4-, 8-, 16-, 32- and 64-bit transmission and 2 mm communication distance. The improvement in performance due to both fast completion detection technique and current sensing signaling was analyzed through design and simulations of the *1of4VmP* and *DualVmP* interconnects. The simulations are carried out for 2-, 4-, 8-, 16-, 32- and 64-bit transmission and 2 mm long links. The distance between the pipeline registers is determined in such away that the interconnect achieves optimal throughput, that is, when the sum of encoder, driver and wire delays (both forward and backward) becomes greater than the delay of completion detection at the pipeline stage. It is 0.4 mm for both 1-of-4 and dual-rail encoded transmission as determined from the simulation.

5.5.3 Performance Analysis

Latency and throughput are used as main parameters to analyze the performance of all the interconnects. The forward latency of the *PMCmFCD* and *DualdiffFCD* links are shown in Fig. 5.10. This latency is the sum of the encoder, driver, wire, receiver, and completion detection delays. The *DualdiffFCD* link has a smaller forward latency and the difference in latency between the two links becomes larger with the increasing wire length. This is because the dual-rail link uses differential current sensing signaling which is faster than the single-ended current sensing signaling used in *PMCmFCD*. The backward latency is the same for both links since they use the same acknowledgment transmission interconnect. The backward latency is smaller than the forward latency since there is no encoding, decoding or completion detection involved.

The throughput of the *PMCmFCD* and *DualdiffFCD* links is analyzed for 2-bit transmission because it is the smallest possible transmission for a 1-of-4 encoded channel. The result is shown in Fig. 5.11. The *PMCmFCD* and *DualdiffFCD* links achieve throughputs of 6.920 Gbps and 7.936 Gbps, respectively for the 1 mm communication distance. When the communication distance increases to 3 mm their

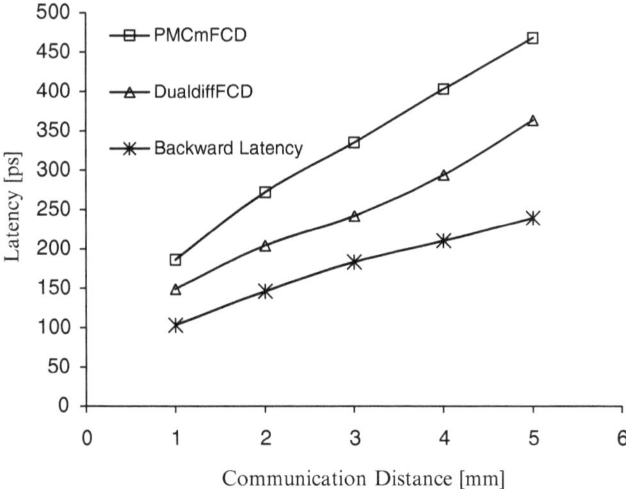

Fig. 5.10 Latency of links with high-speed completion detection

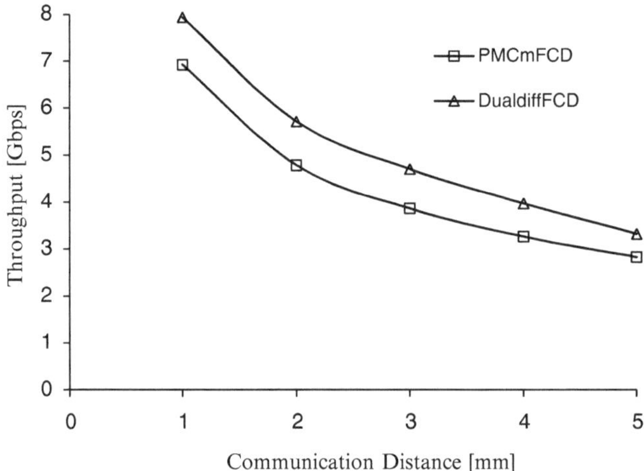

Fig. 5.11 Throughput of links with high-speed completion detection

throughputs become 3.788 Gbps and 4.705 Gbps, respectively, which indicates 45% and 41% reduction in throughput, compared to the 1 mm case.

The major contribution of the high speed completion detection scheme comes into picture when the transmission bit width increases. The throughput of the two-phase 1-of-4 encoded links (*PMCmFCD*, *PMCm* and *1-of4VmP*) is shown in Fig. 5.12 for different bit widths and the wire length of 2 mm. The difference in throughput between *PMCmFCD* and the other two becomes larger when the bit width increases, showing the advantage of the proposed high speed completion

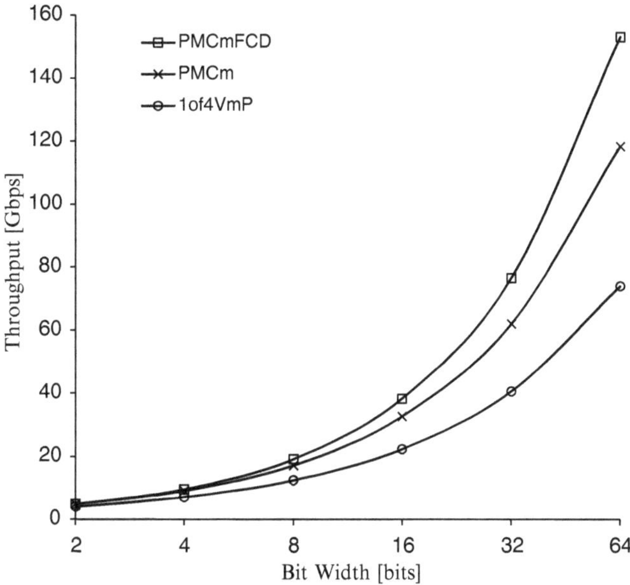

Fig. 5.12 Throughput of two-phase 1-of-4 encoded 2 mm interconnects

detection technique. For example, in the 64-bit case, throughput of *PMCmFCD* becomes 1.29 times that of *PMCm* and 2.07 times that of *1of4VmP*. The throughput gap between *PMCmFCD* and *PMCm* shows the performance improvement due to the high speed completion detection, because the difference between the two links is only in the implementation of completion detection. The difference between *PMCmFCD* and *1of4VmP* reveals the advantage of current sensing signaling along with the proposed completion detection technique.

In the case of two-phase dual-rail encoded interconnects, the *DualdiffFCD* interconnect achieves a higher throughput than the others as expected and the gap increases with the bit width. Its throughput for 64-bit transmission becomes 1.54 and 1.72 times that of *Dualdiff* and *DualVmP*, respectively. The throughput of these interconnects is shown in Fig. 5.13. It can be seen that the current sensing interconnect with the presented fast completion detection technique is a better alternative. It improves the performance of the delay-insensitive links significantly, especially for wider bit-parallel transmissions, compared to the conventional implementation based on pipelined voltage-mode signaling.

5.5.4 Power Analysis

The average total power consumption of the three 1-of-4 encoded 2 mm interconnects for different bit widths is shown in Fig. 5.14. The *PMCm* interconnect

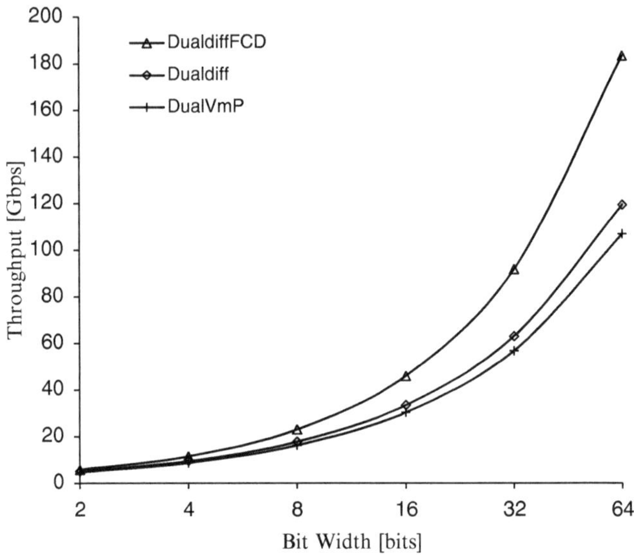

Fig. 5.13 Throughput of two-phase dual-rail encoded 2 mm interconnects

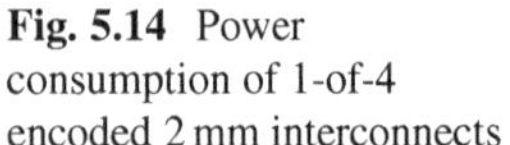
Fig. 5.14 Power consumption of 1-of-4 encoded 2 mm interconnects

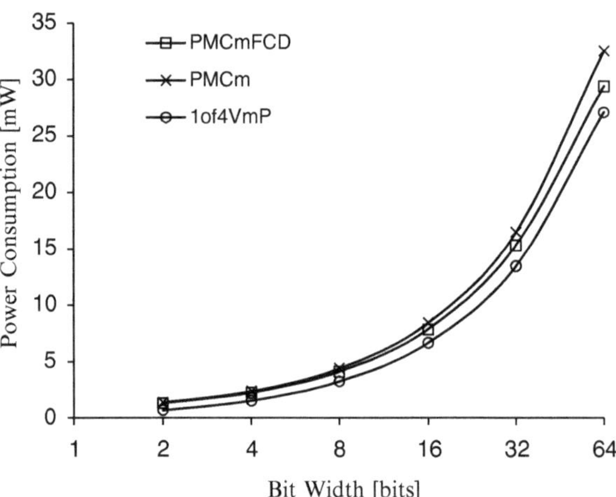

consumes the highest power compared to the other two. For example, in 64-bit transmission *PMCm* consumes 10.6% and 20.1% more power than that of *PMCmFCD* and *1of4VmP*, respectively. Similarly, the dual-rail encoded interconnect with conventional completion detection, *Dualdiff*, consumes the highest power compared to the other two. It consumes 19.5% to 26.2% more power than that of *DualdiffFCD*. For 64-bit transmission, the *Dualdiff* interconnect consumes 32.2% more power than that of the *DualVmP*. *DualdiffFCD* consumes slightly more power

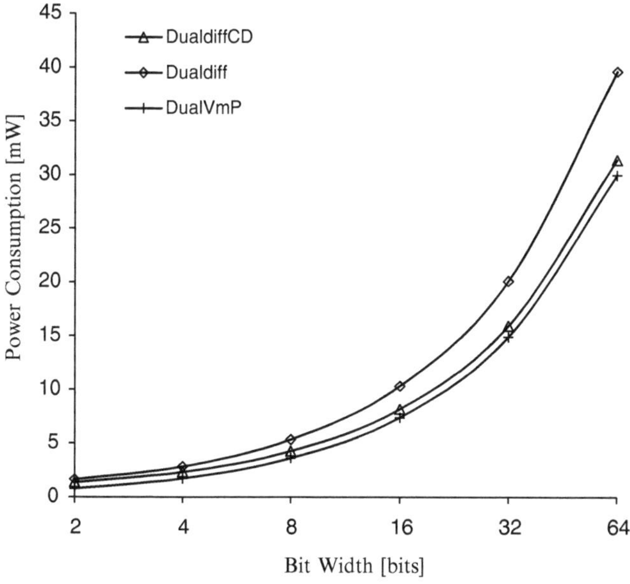

Fig. 5.15 Power consumption of dual-rail 2 mm encoded interconnects

than *DualVmP*, for instance, for 32- and 64-bit transmissions it consumes 6.9% and 4.8% more power compared to *DualVmP* power consumption. The power consumption of the three dual-rail encoded interconnects is shown in Fig. 5.15.

In order to determine the power efficiency of these links, the energy per bit dissipated by the individual interconnects is examined. The energy per bit metric combines the power consumption and performance figures, which allows to judge the efficiency of these interconnects in a reliable manner. The energy per bit of the 1-of-4 and dual-rail encoded links with 2 mm wire length and 2–64-bit transmission is shown in Figs. 5.16 and 5.17, respectively. For 2- and 4-bit transmission, the conventional links, *1of4VmP* and *DualVmP* have a better power efficiency. When the bit-width is more than 4 bits, *DualdiffFCD* is the most energy efficient and closely followed by the *PMCmFCD* link. Regarding the two voltage-mode reference links, energy efficiency deteriorates with increase in bit width because the completion detection circuit becomes an increasingly large tree of logic elements causing a significant power overhead. Hence, the current sensing delay-insensitive links with the proposed completion detection technique not only boost the performance of communication but also improve the power efficiency. This makes them most appropriate global interconnects in nanometer technologies where delay variations and power consumption are the major concerns.

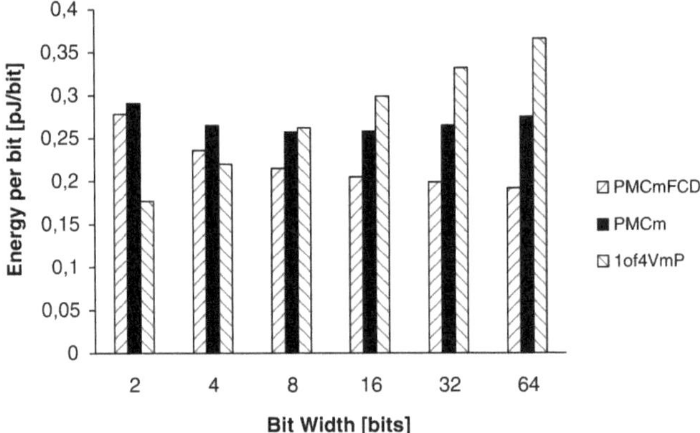

Fig. 5.16 Energy per bit dissipation of 1-of-4 encoded 2 mm links

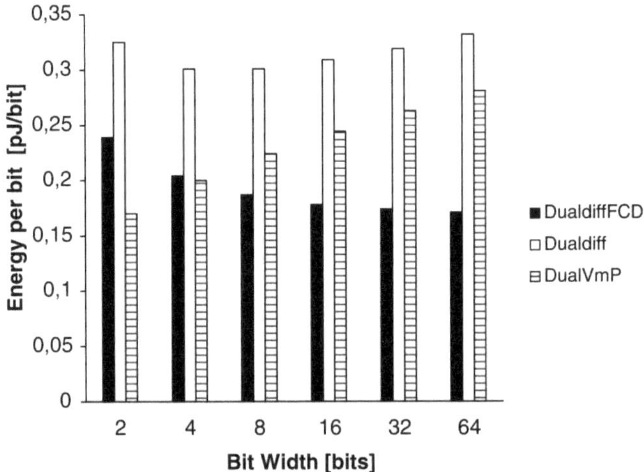

Fig. 5.17 Energy per bit dissipation of dual-rail encoded 2 mm links

5.5.5 *Noise Analysis*

To assess the impact of noise on the proposed links, noises of 150 mV amplitude with variable frequencies were introduced into the driver circuit and simulations were carried out. In case of *PMCmFCD* interconnect with 150 mV and –150 mV noise, its receiver's input current decreases by 36 μA and increases by 30 μA, respectively. In addition, the reliability of *PMCmFCD* interconnect were examined under power supply noise of 10%VDD. Despite these noises, the *PMCmFCD* interconnect outputs data and data validity indicator signals correctly because the wire output current variation was within the allocated current margin. The effect of noise on *DualdiffFCD* interconnect was also examined using the same noise types as above. The 150 mV noise coupled in the *DualdiffFCD*'s driver causes wire output current to decrease by 21 μA and increase by 24 μA. The 10%VDD power supply noise makes the *DualdiffFCD*'s wire output current to decrease and increase by 29 μA and 33 μA, respectively. The DualdiffFCD interconnect also outputs the data and data validity signals reliably despite the presence of these noises.

5.5.6 Post-Layout Simulation

In addition to the above schematic based simulations, post-layout simulations were also carried out. The main purpose of the post-layout simulation is to verify the correctness of the presented design techniques as well as the overall interconnects. In case of *PMCmFCD* interconnect post-layout simulation for 1.8 mm long 32-bit data transmission was done. The layout of this interconnect is shown in Fig. 5.18. The data along with its validity indicator signal (*Reqout*) was received and decoded correctly at the receiver side. Also, the acknowledgment signal transmission was performed successfully. The post-layout simulation of *DualdiffFCD* interconnect was done for 16-bit and 1 mm long communication distance. Its layout is shown in Fig. 5.19. The layout of *DualdiffFCD* link is more compact because effort is made to fit this interconnect in the smallest possible area.

5.5.7 Area Comparison

Minimizing the area overhead of any module in a chip has been one of the concerns especially in the current GSI era. The link area, which consists of global wires and their signaling circuits, takes a significant portion of the total chip area. In this regard, it is wise to examine the areas of the proposed links and compare them with the conventionally implemented links. The areas of 2 mm long 64-bit wide interconnects calculated from the link schematics, are presented in Table 5.1. The two current sensing links with high speed completion detection, *PMCmFCD* and *DualdiffFCD*, require a smaller silicon and metal area than their voltage-mode counterparts. The current sensing links require only 82% of the wiring area of the pipelined links. The reason for this is that in pipelined links one acknowledgment wire is needed for every 2-bit transmission to minimize the delay incurred due to completion detection at each pipeline stage. Only one acknowledgment wire is required for the whole link in the current sensing links since completion detection needs to be performed only at the receiver. The current sensing links also require less silicon area compared to the pipelined voltage-mode ones because there are no area consuming pipeline stages.

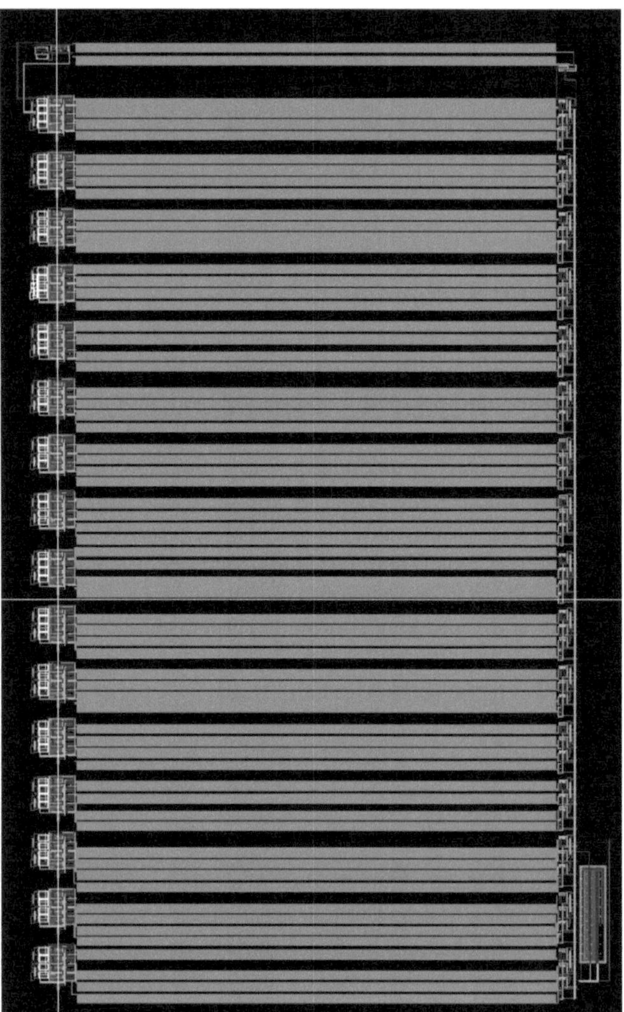

Fig. 5.18 Layout of 32-bit *PMCmFCD* interconnect

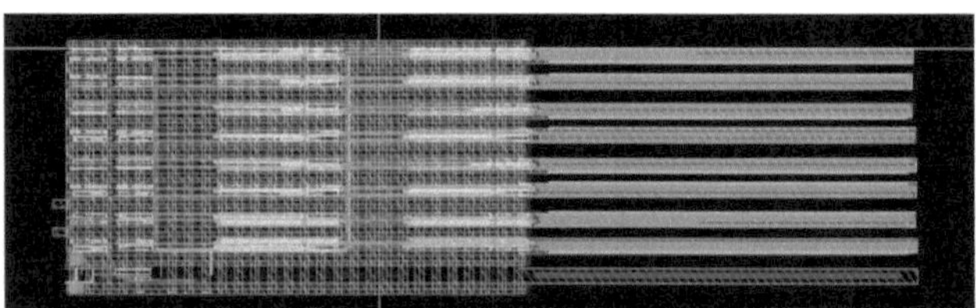

Fig. 5.19 Layout of 16-bit *DualdiffFCD* interconnect

Table 5.1 Area comparison of interconnects

Link name	Active area [μm^2]	Wiring area [μm^2]
PMCmFCD	508	55,020
1of4VmP	704	66,780
DualdiffFCD	599	55,020
DualVmP	662	66,780

5.6 Chapter Summary

A performance boosting technique by using a high speed completion detection circuit for delay-insensitive on-chip interconnects has been presented. Delay-insensitive data transfer is a necessity for global links in nanometer-scale technologies where delay variations are inevitable. The performance and power analysis shows that using the presented high speed completion detection technique improves throughput and power efficiency compared to the conventional implementations. It also requires less silicon and metal area. Therefore, using the presented technique leads to a realization of high throughput and power efficient global communication.

Chapter 6
Energy Efficient Semi-Serial Interconnect

Design and analysis of a high-throughput self-timed serial on-chip communication link is presented. Using fully bit-parallel interconnects that are presented in the previous two chapters for long-range communication links incurs considerable area overhead, routing difficulty, severe crosstalk noise and significant leakage power, making serial links a better alternative. The analysis between parallel and serial links in [108] and [109] shows the tradeoff between link length, latency, dynamic and leakage power as well as active and wiring area. For a given throughput the serial link is always preferable in terms of wiring area and incurs less routing congestion than parallel links. The serial link also takes smaller active area and consumes less leakage and dynamic power than the parallel link for long global communication [109].

In source-synchronous serial communication a clock is injected into the data stream at the transmitting side and the clock signal is recovered at the receiver side. Such clock-data recovery (CDR) circuits often require a power-hungry PLL, which may also take a considerable amount of time to converge on the proper clock frequency and phase at the beginning of each transmission. If the receiver and the transmitter operate in different clock domains, the transaction must also be synchronized at both ends, incurring additional delay and power consumption. One such link is presented in [108], it uses wave-pipelined multiplexed routing technique and its performance is limited by the clock skew and delay variations. In [34], circuits that had originally been used for off-chip communications [84, 85] were adopted to design a serial on-chip link. It uses output multiplexed transmitter architecture due to its ability to deliver better performance than input multiplexing. However, this comes at the expense of much higher output capacitance that grows linearly with the bit-width. Both transmitter and receiver use multi-phase DLL circuits and clock calibration is required at the receiver side. A prototype chip has been fabricated in 180 nm CMOS technology and a 3 mm long link has achieved a throughput of 8 Gbps. Total power consumption or energy per bit of this link is not reported. Another high-speed serial link was presented in [82], where the

E.E. Nigussie, *Variation Tolerant On-Chip Interconnects*, Analog Circuits and Signal Processing, DOI 10.1007/978-1-4614-0131-5_6,

serializer/deserializer are based on a chain of MUXes. The link is single-ended and employs wave-pipelining. As a timing reference, constant delay elements are used instead of clock. Furthermore, the operation is based on the assumption that the introduced unit delays for the serializer and deserializer are the same. However, getting the same delay is almost impossible in the sub-100 nanometer CMOS technology due to considerable PVT variations, which in turn makes the reliability of communication using this approach questionable. A test chip was manufactured using 180 nm CMOS technology and the measured throughput was 3 Gbps. The power consumption or energy per bit dissipation of this link is not reported.

The energy efficiency advantage of equalized interconnects for long-range transmission has been demonstrated in [163–167]. In [164], a capacitively driven interconnect is presented. The capacitor pre-emphasizes signal transitions that leads to a decrease in wire delay and reduces the driver load. The measured result from a prototype chip in 180 nm technology shows 0.28 pJ/bit/mm energy dissipation for a bus running at 1 GHz. An equalized interconnect which uses capacitive pre-emphasized transmitter and decision feedback equalizer at the receiver in 90 nm CMOS process has been proposed in [165]. At a data rate of 2 Gb/s, the transceiver of this interconnect consumes 0.28 pJ/bit. In [166], a transceiver consisting of a nonlinear charge injecting transmit filter and a sampling receiver with a transimpedance pre-amplifier is proposed. The measured energy dissipation of this transceiver manufactured in 90 nm technology is 356 fJ/bit at 4 Gb/s data transmission in 10 mm long interconnect. Further performance improvement over interconnects presented in [164, 165] has been reported by using a transmit-side adaptive FIR filter and a clockless receiver in a capacitively driven pulse-mode wire [167]. This interconnect achieves 4.9 Gb/s and dissipates 0.34 pJ/bit energy for 5 mm link manufactured in 90 nm technology. The power saving benefit of equalized interconnects in the long-range channels of low-diameter on-chip networks has been demonstrated in [163].

An alternative approach is to use self-timed communication which employs handshake instead of clocks. This enables robustness to delay variations through the use of delay-insensitive encoding and data transfer. A high data rate asynchronous bit-serial link for long-range on-chip communication is presented in [83] and its improved version in [168]. It uses two-phase LEDR data encoding, fast asynchronous shift registers for both serializer and deserializer and wave-pipelined differential current-mode signaling. Due to direct integration of LEDR and differential signaling, this communication requires four wires per one link, increasing the required area and energy per bit of the bit-serial link. It achieves one-gate delay data bit cycle, that is, 67 Gbps throughput in 65 nm CMOS technology. The reported total power consumption of a 7 mm link when transferring 16-bit word at 20% utilization is 35 mW [168].

In today's SoCs power dissipation is a major design constraint that limits battery life and reliability, emphasizing the need for low-power on-chip communication links, which is the major motivation of this work. The argument is that it is possible to achieve 67 Gbps throughput using one bit-serial link by restricting the data cycle to gate delay. However, the power dissipation of such a link is unacceptably high [83]. On the other hand, it is possible to achieve the same or even higher throughput with smaller power consumption by having few bit-serial links in parallel which are designed from simple customized circuits and combination of techniques. Thus, the serial link presented in this chapter adopts the low-power approach.

This chapter is organized as follows. In the next section the need for long-range on-chip communication link in NoC is discussed. The proposed serial link communication protocol and detailed design of its circuits (serializer, deserializer, driver, receiver and data validity decoder) are presented in Sect. 6.2. Spice-level simulation results and analysis of the link performance, power, energy and area are discussed in Sect. 6.3. Comparison of fully bit-parallel, bit-serial and semi-serial links in terms of performance, energy and area is presented in Sect. 6.4. Finally, Sect. 6.5 presents the summary of this Chapter.

6.1 Long-Range Link in NoC

Most of NoC research has been focused on microarchitecture improvement and routing algorithms. However, selecting an appropriate topology is also one of the most critical decisions because it bounds critical performance metrics such as the network's zero-load latency and its capacity [71] and affect energy efficiency. The most common NoC topologies that have been used so far are 2-D mesh [68]. For example, the 80-node Intel's TeraFLOPS [69], the 64-node chip multiprocessor from Tilera [66], and the 167-processor computational platform [64] are implemented using 2-D mesh network. These networks have short wires in the architecture, but they have long network diameter. This causes energy inefficiency because of extra hops and furthermore consumes area. For instance, the 16-tile MIT RAW on-chip network consumes 36% of total chip power [70] and Intel's TeraFLOPS link and routers consume 28% of the tile power [69].

There are NoC topologies which require long-range links such as torus (not folded-torus) [71], flattened butterfly [72], Spidergon [74], and concentrated mesh with replicated subnetworks and express channels [73]. It has been shown in [72] that using flattened butterfly with high-radix router offers lower latency and power consumption than 2-D mesh. That is, the latency and power consumption of flattened butterfly has been reduced by 28% and 38%, respectively compared to 2D-mesh. In [73], detailed area and energy models for different on-chip networks have been developed and their design tradeoffs are analyzed. It has been shown that concentrated mesh with replicated subnetworks and express channels provides a 24% improvement in area efficiency and a 48% improvement in energy efficiency over other topologies (mesh, folded-torus, concentrated mesh, fat-tree and tapered fat-tree). The express channel contributes 23% area and 38% energy efficiency. The area overhead is negligible because the express channels are routed over processor tiles in otherwise unused metal tracks and use preexisting router ports. The significant energy efficiency improvement is due to the decrease in completion

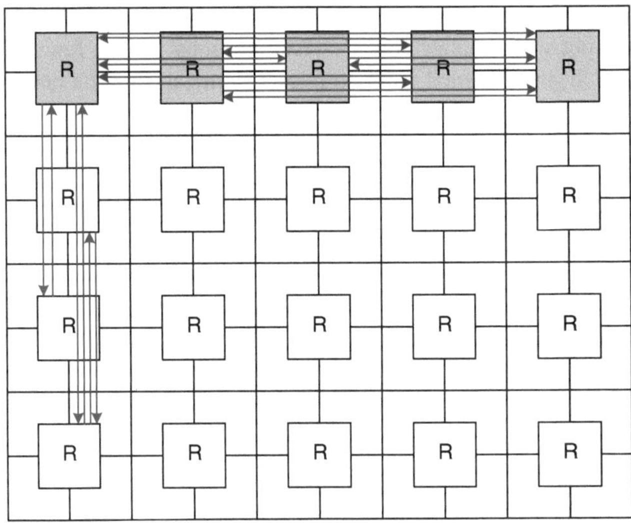

Fig. 6.1 80 nodes flattened butterfly topology

time and due to the increased routing efficiency. That is, it is more efficient to route packets over express channels than through intermediate routers. Both flattened butterfly and concentrated mesh with express channels require long-range links which span more than one tile (see Figs. 6.1 and 6.2). There are also other NoC topologies such as Spidergon [74, 75] and torus which are efficient and require longer links (Fig. 6.3). Furthermore, in [76] topologies with fewer hops and longer channels have been proposed as promising solutions for energy and area efficient on-chip interconnection networks. It has also been demonstrated that adding few additional long-range links in a mesh network reduces the average packet latency significantly and improves the achievable throughput substantially [78]. Experiments involving real data traffic from telecom applications shows that the insertion of long-range links provides 36.3% improvement in critical traffic load, and 61.4% reduction in packet latency [78]. All these show the importance and need for high-speed and low-power long-range link, where its length spans two or more tiles.

Using fully bit-parallel communication in long-range NoC links that traverse two or more tiles becomes costly because it requires larger chip area, introduce routing difficulties, severe crosstalk noise and considerable leakage power (due to large driver/receiver to communicate through long lossy wires). Most of these issues can be addressed by using a high-speed serial link. Therefore, a high-throughput and low-power long-range serial on-chip link that can be used in NoC topologies which require long-range links inherently or when customized is presented. This in turn increases the overall network throughput and decreases the power consumption besides minimizing traffic congestion.

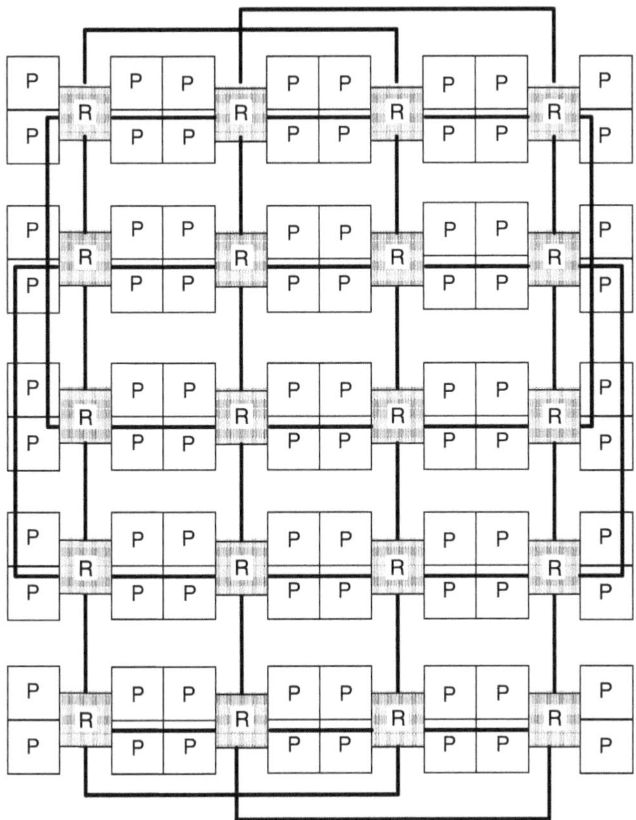

Fig. 6.2 80 nodes concentrated mesh with express channel

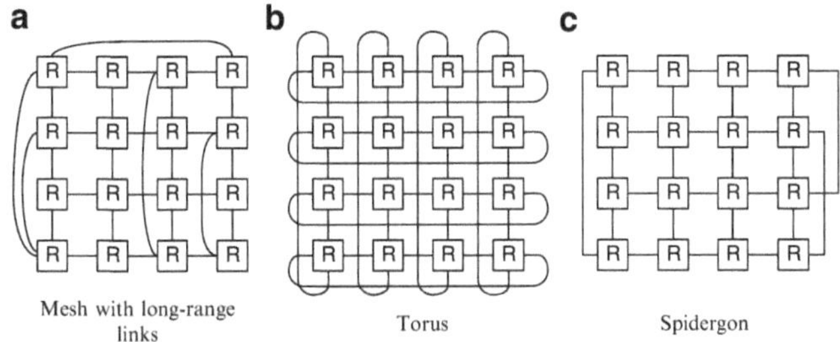

Fig. 6.3 16 nodes NoC topologies requiring long-range links

6.2 High-Throughput Serial On-Chip Interconnect

A high-throughput and low-power serial on-chip communication link employing integration of pulse dual-rail data encoding, wave-pipelining, pulse signaling and differential current-mode signaling is presented. Two-phase pulse dual-rail encoding is performed at low cost using two *AND* gates, one for data bit "1" and the other for "0". This encoding enables usage of pulse signaling along with differential signaling directly. Furthermore, both the latency and the power consumption are reduced because data decoding logic is not needed at the receiver. The ability to detect each bit through pulse signaling in the wave-pipelined communication makes the link delay-insensitive and also enables acknowledging the transmission per word instead of per bit, improving throughput and saving energy.

In the presented serial link customized circuits and logics for serialization/deserialization and fastest possible stoppable local clock in the serializer are implemented. High-speed differential pulse current-mode signaling circuits are also designed. In addition, fast and robust data validity decoding circuits are designed. With this, one serial link achieves 9.09 Gbps throughput. The serial link consists of serializer and deserializer, dual-rail encoder, driver, receiver and data validity decoder, as shown in Fig. 6.4. In the subsequent sections, the communication protocol, design details of the link circuits and the signaling technique will be discussed.

6.2.1 Communication Protocol

Similarly with the other interconnects, presented in this thesis, it is assumed that the sender and receiver modules have two-phase bundled-data interface. As soon as there is a request from the sender module which informs the data to be sent are ready and stable, the data will be loaded into the shift register. In addition to the data, the *Stop* bit is also loaded which will be used to stop the shifting in the deserializer without the need for additional control logic such as data bit counter. The locally generated clock starts running after parallel data loading is completed. It is used for data shifting and dual-rail data encoding. It is a stoppable clock that runs only

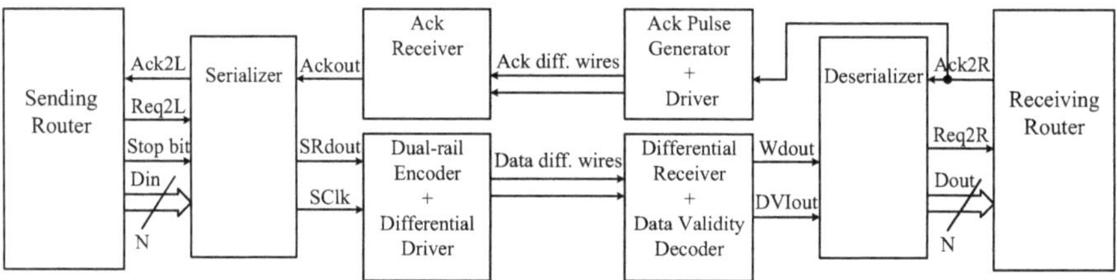

Fig. 6.4 High-throughput serial on-chip communication link

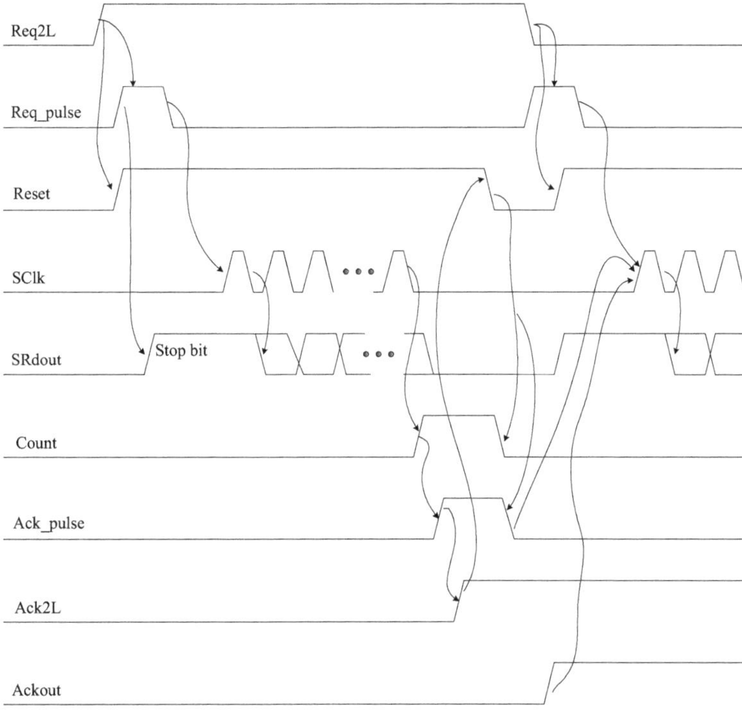

Fig. 6.5 Serializer communication protocol

when there is data in the shift register to be transmitted and stopped at all other time, saving the communication power significantly. Data is shifted at the negative edge of the clock and encoded when the clock is in *high* state. The counter counts at the negative edge of the clock and signals the completion of data shifting when it reaches the maximum count value, which in turn stops the clock.

Dual-rail and differential pulse current-mode signaling is used for data transmission through the wire. Acknowledgment is sent per word instead of per bit thanks to the devised delay-insensitive wave-pipelining in the wire. In the receiving side, the transmitted data is retrieved directly from the receiver without the need for data decoding logic. The extracted data validity indicator is used as a clock for shifting the data in the deserializer. Shifting is performed at both edges of the data validity indicator signal. The arrival of *Stop* bit at the last flip-flop of the deserializer indicates that shifting is completed and the data are ready for parallel bit out. At this point, request to the receiving router will be sent. The deserializer shift register will be cleared when an acknowledgment from the data receiving module is received.

The overall communication protocol of the serializer is shown as a timing diagram in Fig. 6.5. *Req2L* and *Ack2L* are the two-phase bundled-data request and acknowledgment signals of the sender. *Req_pulse* is used to enable parallel data

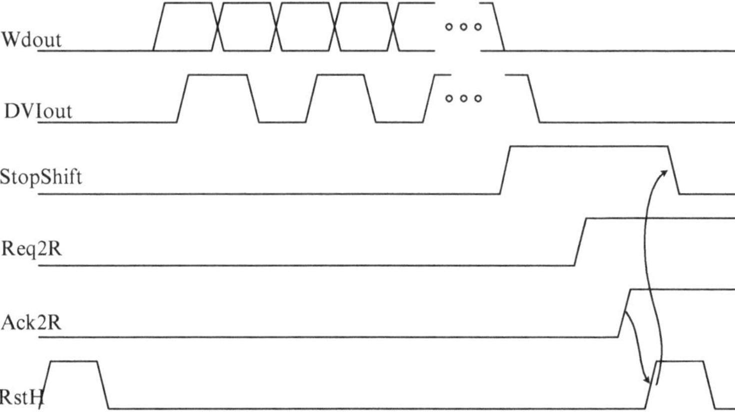

Fig. 6.6 Deserializer communication protocol

loading in the shift register. *SRout* is the serializer's shift register data output and *Clk* is the locally generated clock. *Count* is the counter output signal and becomes *high* when all data are shifted out from the serializer's shift register. *Ack_pulse* is generated from *Count*, and it is used to stop the clock. It also saves communication time between data bursts by allowing data loading to be performed whilst waiting for acknowledgment to arrive from the receiving side. *Reset* is a locally generated signal which is used to reset the counter's registers besides allowing the clock to start running again by putting down the *Ack_pulse* to *low*.

The deserializer consists of a shift register and interfacing circuit between the serial link and receiving router. Its communication protocol is shown in Fig. 6.6. The data receiver output *Wdout* is shifted in the deserializer shift register at both edges of data validity indicator signal, *DVIout*. The shifting process will be stopped when the *Stop* bit reaches to the shift register's last flipflop. *Req2R* and *Ack2R* are the bundled-data interface between the link and receiving block (Fig. 6.4). *RstH* signal resets the deserializer's shift register.

6.2.2 Serializer and Pulse Dual-Rail Encoding

The bit parallel data from the sender is serialized using a novel shift register which uses the locally generated clock to shift the stored data. As shown in Fig. 6.7, the serializer consists of shift register, counter, clock generating circuit and other interfacing elements. The design of the shift register is based on True Single-Phase-Clocked (TSPC) flip-flops [91] and customized to have parallel data loading ability. TSPC is chosen because of its ability to embed logic, parallel data loading in this case, with very little delay overhead. In addition, it has much smaller setup time and propagation delay compared to other dynamic flip-flops, making it the most

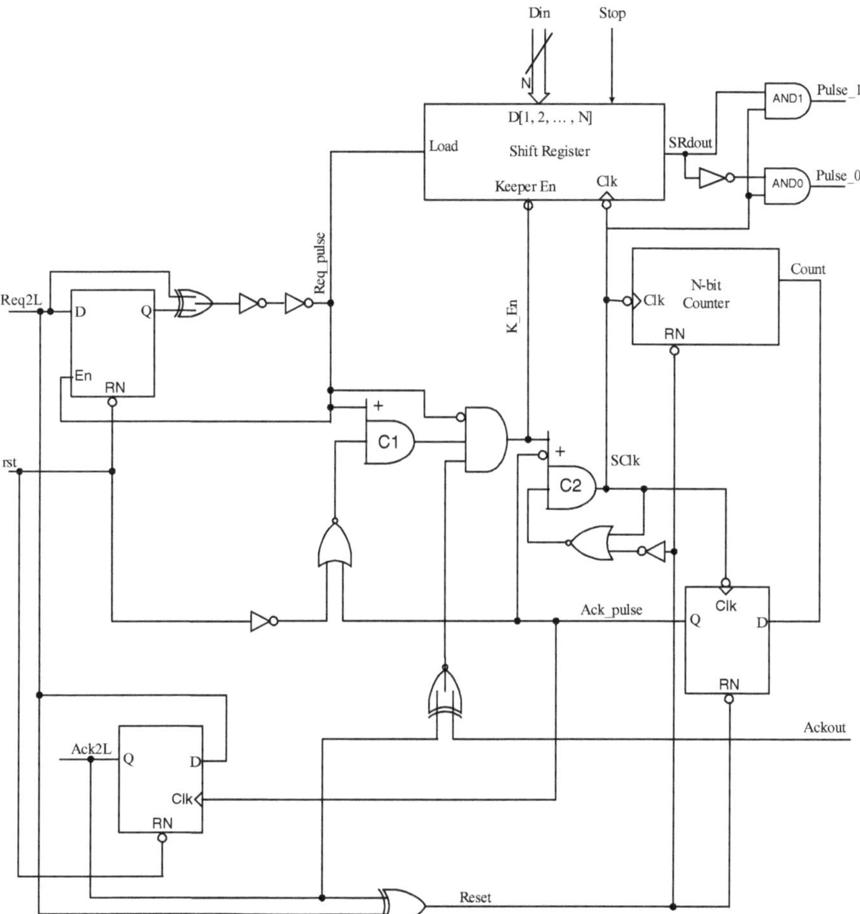

Fig. 6.7 Serializer and pulse dual-rail encoder

suitable to realize high-speed shift registers. The customized TSPC circuit with parallel loading is shown in Fig. 6.8. In the loading phase, transistors *Mns* and *Mnr* are used to load bit "1" and bit "0", respectively and transistor *Mps* decouples *D* from node *L1* (preventing error when *D* is "0" and data to be loaded is "1"). The tri-state weak inverter is used as a keeper for the loaded data. There are two 3-input upper asymmetric C-elements (*C1* and *C2*) in the serializer circuit, shown in Fig. 6.7, that are used to generate the local clock and keeper enable signals. The output of the two NOR gates act as the active-low reset signal for *C1* and *C2*.

One-hot counter is designed from shift register so that its delay becomes equivalent to the data shift register in the serializer. As in the serializer's shift register, the counter shifts its one-hot code at the negative edge of the clock. Its shift register is designed from TSPC flip-flops which are customized to support

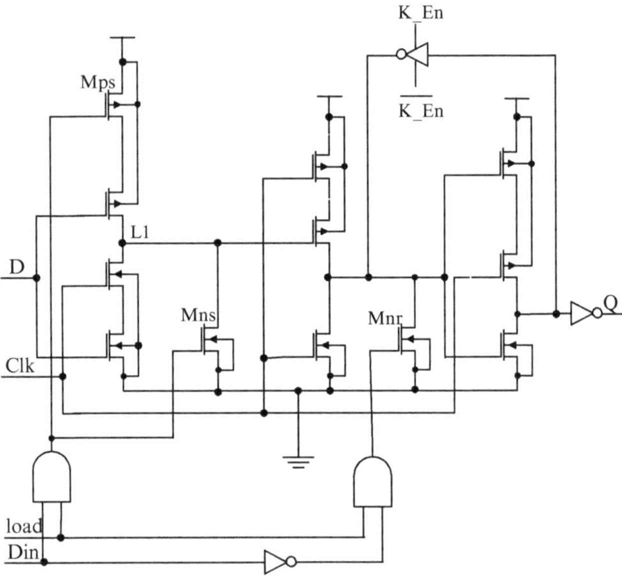

Fig. 6.8 Shift register's TSPC flip-flop with parallel data loading

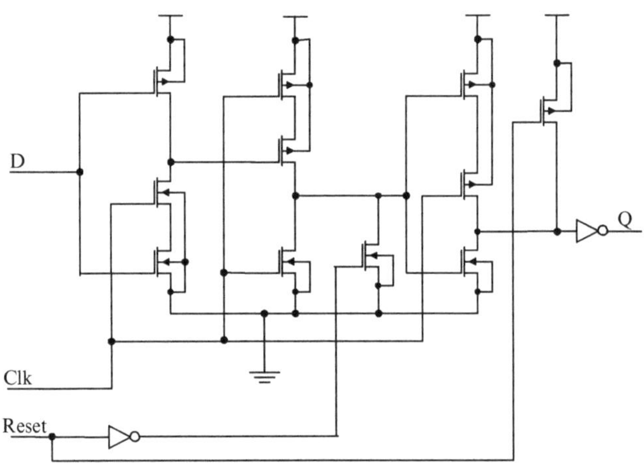

Fig. 6.9 Counter's resettable TSPC flip-flop

active-low reset as shown in Fig. 6.9. For N-bit word counter, N TSPC flip-flops are connected in series and the last flip-flop's output is inverted and feedback to the first flip-flop's input.

As already discussed in Chap. 1, the delay-insensitive data transfer, such as the dual-rail encoded interconnect, is a necessity in global interconnects of a nanometer SoC [92]. The delay-insensitivity makes the data transfer robust, because the sender

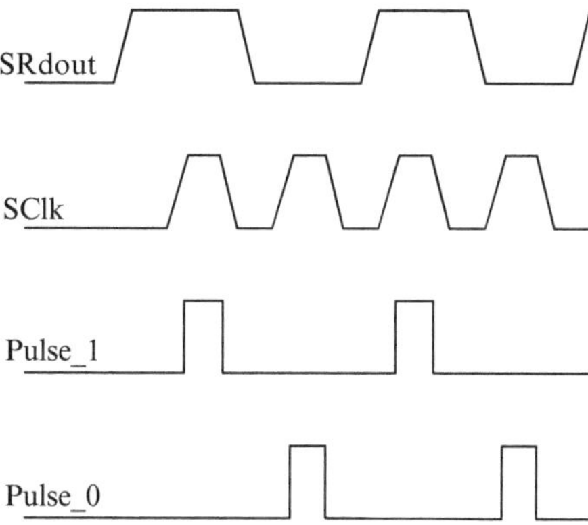

Fig. 6.10 Pulse dual-rail encoder input and output signals

and the receiver modules can communicate reliably regardless of delays in the transceivers and wires. Delay-insensitive data encoding technique requires $2N$ wires to transmit N-bit data. Pulse dual-rail encoding, where the presence of a new valid bit is represented by a pulse instead of voltage transitions or levels, is formulated and used in the presented serial link. This encoding enables straightforward use of pulse signaling. Furthermore, it has simpler and faster encoding/decoding logic when it is used along with differential signaling than the transition based protocols. When the clock is *high*, the dual-rail encoder, shown in Fig. 6.7, encodes each bit into (P, NP) pair depending on its value. For example, when the output of the shift register is bit "1", and the clock is *high*, there is a pulse at the output of *AND1* and no pulse at the output of *AND0*, as shown in Fig. 6.10. Since there is no pulse in both wires between transmission of two consecutive bits, the receiver is able to detect each bit. That is, each bit can propagate at its own speed and can be detected reliably at the receiver regardless of the propagation delay variations.

6.2.3 *High-Speed Differential Pulse Current-Mode Signaling*

In pulse signaling only a small portion of the wire is charged during pulse propagation, significantly reducing the amount of capacitance need to be charged and hence, saving considerable amount of power over level-based signaling. It has been shown that the use of pulse signaling can save up to 50% of energy compared to level-based signaling with repeater insertion [93]. Furthermore, it has been demonstrated through analytical models that more than 70% power saving could be achieved by combining pulse signaling with wave-pipelining technique without penalties of data throughput [94]. Since the main goal of this work is to achieve both high-speed global communication and low-power consumption, pulse signaling along with wave-pipelining is employed. In addition, differential current-mode signaling is used because of its high performance, better energy efficiency and noise immunity features [39–42]. Integration of dual-rail encoding and differential signaling has been realized using only two wires per link instead of four (two for dual-rail and two for differential signaling). This further reduces both power and required area of the link.

In addition to power saving, pulse current-mode signaling mitigates the effect of dispersion due to its return-to-zero signaling scheme in which sharp current pulses are used to transmit data and receiver termination is employed. To make use of these promising advantages, the wires need to be modeled with consideration of the lossy on-chip environment. Wider and thicker wires with larger spacing than the minimum is preferred to ease attenuation and preserve pulse integrity. This can be realized with smaller area overhead in a serial link than in parallel links.

In this link, the driver sinks a narrow current pulse from one of the wires and no current from the other wire for every bit transmission. The driver sinks no current from both wires during bit spacer transmission (delay-insensitive encoding) and also between data bursts. The receiver detects the voltage difference between the two wires and amplifies it in order to retrieve the value of the transmitted bit. Bit level data validity detection is carried out by sensing the amount of current in the wires. Driver, receiver and data validity decoder circuits design and operation are described in detail in the following sections.

6.2.3.1 Driver with Pre-emphasis

A source-coupled differential current-steering driver, shown in Fig. 6.11, along with pre-emphasis is designed and used. It is fast because it has an extremely sharp transient response. The driver also has an advantage of reducing the AC component of the power supply noise because the circuit draws constant current from the supply. This driver is naturally suited to drive a balanced differential pair of wires. The complementary outputs of the driver are attached to the two wires.

Depending on the output of the dual-rail encoder, current will be steered in one of the wires. When bit "1" is transmitted, *Pulse_1* sets *Mnp1* in conducting phase. This in turn steers a current pulse in *wire1* and no current in *wire0*. When bit "0" is transmitted, the current pulse passes through *Mnp0*, which in turn steers the current pulse in *wire0* and no current in *wire1*. The amount of current in the wire is a function of driver's current source, wire impedance and termination load. In order to get the same amount of current at the end of the wire irrespective of its length, the sizes of driver's transistors have to be adjusted accordingly.

The change in wire impedance due to signal frequency change causes signal distortion. To solve this frequency-dependent signal degradation, driver pre-emphasis technique is employed. Driver pre-emphasis compensates the channel high frequency loss by either emphasizing the high frequency signal component or attenuating the low frequency components to transmit an equalized signal to the receiver input. Equalization at the driver side is easier to implement for non-variant channels like on-chip interconnects [162]. In the proposed serial link, when signaling after an idle period, for example between two data bursts or after a spacer, the characteristic impedance is high which may cause the transmitted pulse to be

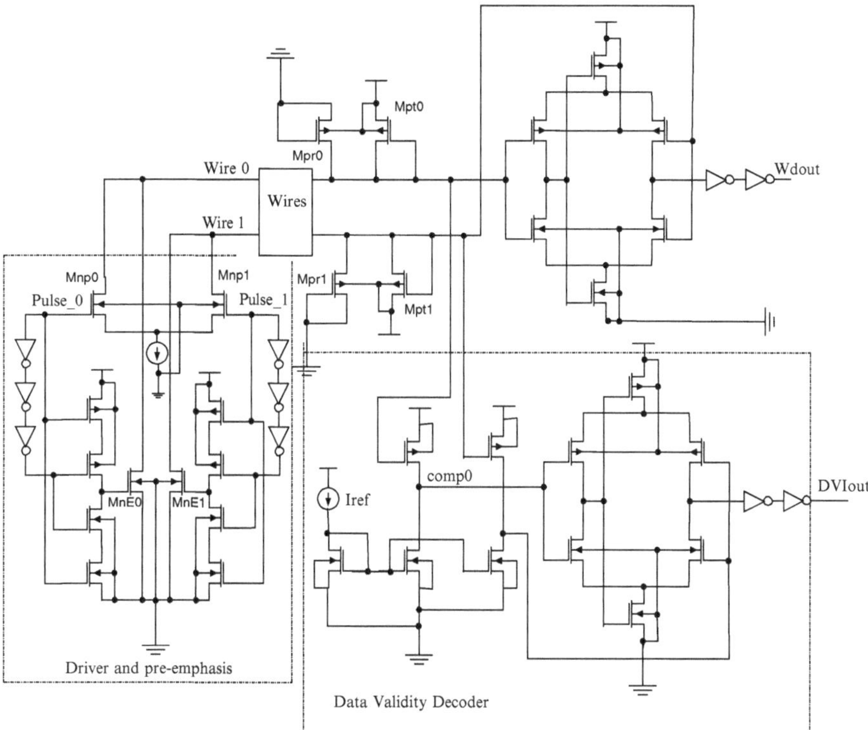

Fig. 6.11 Driver, receiver and data validity decoder of serial link

distorted. The pre-emphasis inverters and pull-down/pull-up transistors constitute inertial delays and control a variable load on the driver output. During the pre-emphasis period, transistors *MnE1* and *MnE0* sink additional current in order to provide extra driving capacity.

6.2.3.2 Receiver and Data Validity Decoder

The termination load and receiver design is shown in Fig. 6.11. Diode connected *Mpt0* and *Mpt1* transistors are used as termination load. In addition to termination, they are also used to mirror the wire current which will be needed to decode out data validity indicator. The transconductance of these transistors has been regulated through the use of *Mpr0* and *Mpr1*. The receiver needs to have high common-mode noise rejection capability in order to take full advantage of differential signaling. Due to this, a high-speed self-biased differential amplifier is used. The differential amplifier used in this design has less sensitivity to process, temperature and supply voltage variations. It operates at high speed because its output switching currents are significantly greater than its quiescent current. Furthermore, the adopted amplifier has higher differential-mode gain than conventional amplifiers and a large common-mode input range because its bias condition adjusts itself to accommodate the input swing [91, 95].

In delay-insensitive transmission, decoding of data and data validity indicator at the receiving end is necessary. The transmitted data is received and decoded out directly in the receiver without the need for separate data decoding logic. This is due to the novel integration of pulse and differential signaling. The remaining issue is data validity indicator, which will also be used as a clock to shift the data bit in the deserializer. From the encoding, it is known that there will be current only in one of the wires when there is valid bit transmission and no current in both wires between two consecutive bit transmissions. Each wire's current is compared with a reference current using a current comparator and the output of the two current comparators is fed to a differential amplifier. The output of the differential amplifier is the data validity indicator (*DVIout*). This way of completion detection makes the communication robust to both delay variations and noise because it takes into account both wires and the used differential amplifier, which has a high common-mode noise rejection ratio. Both edges of *DVIout* signal indicate the availability of valid and new data at the receiver output. The circuit of the data validity decoder is shown in Fig. 6.11.

6.2.4 Deserializer

The deserializer consists of a shift register and interfacing circuit (between the receiving module and the deserializer) as shown in Fig. 6.12. In shift register data is shifted out at both edges of the *DVIout* signal. The shift register is designed from double-edge-triggered flip-flops. This flip-flop is designed by tying together the outputs of a negative and a positive edge-triggered TSPC flip-flops, obtaining multiplexer function for free. It stores dynamically during opposite clock phases and drives its output actively on both clock edges. The circuit of a double-edge-triggered flip-flop is shown in Fig. 6.13. *Mnrs1* and *Mnrs2* transistors are used for resetting the flip-flop. When *Stop* bit reaches the last FF, the shifting will be stopped and data can be read out in parallel. The bundled-data two-phase request signal for parallel data receiving module is generated using a D-FF as shown in the interfacing circuit (Fig. 6.12). As soon as an acknowledgment is received, the deserializer's shift register will be cleared (resetted).

6.2.5 Acknowledgment Transmission

As already discussed, acknowledgment is sent from the receiver per word instead of per bit. The same signaling technique as in data transmission is used except that there is no wave-pipelining, as it is not necessary (there is only one bit to transmit

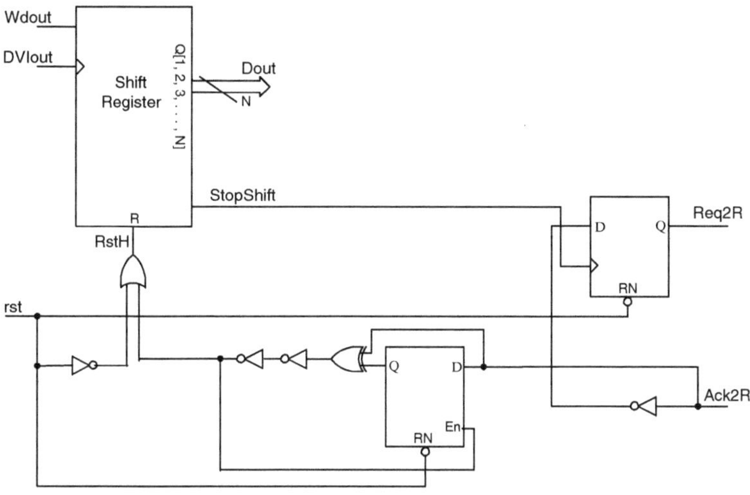

Fig. 6.12 Deserializer circuit

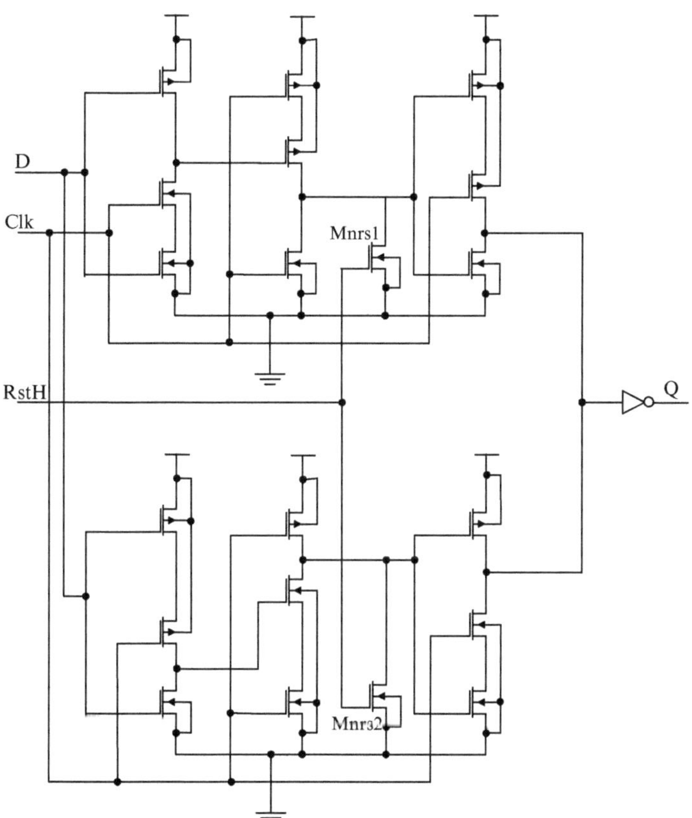

Fig. 6.13 Double edge-triggered TSPC flip-flop

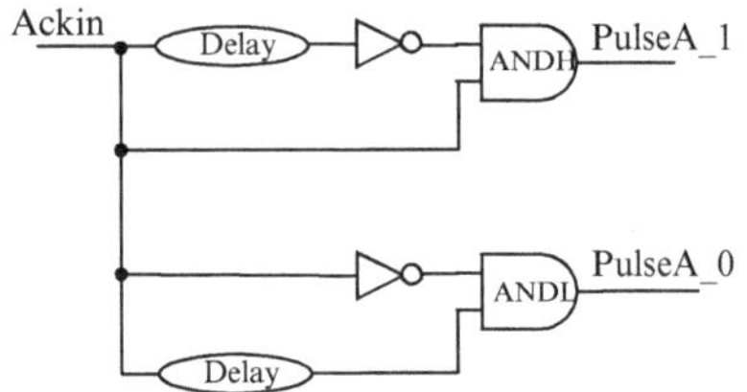

Fig. 6.14 Pulse generator for acknowledgment signal transmission

at a time). Since the receiver has a two-phase handshaking interface, a pulse is generated and transmitted at each transition edge of the acknowledgment signal. The pulse generator circuit, shown in Fig. 6.14, generates a pulse for low-to-high and high-to-low transitions. Its driver and receiver circuits are similar with the data transmission circuit, shown in Fig. 6.11, but the wires are narrower. This is due to the fact that the performance of acknowledgment transmission is not detrimental for the throughput of the link because the transmission takes place once per word.

6.3 Simulation Results and Analysis

The performance, power consumption, and energy per bit of the presented serial link is discussed in this section. In addition to the bit-serial link, semi-serial (two, four and eight bit-serial) links are simulated and analyzed. In the bit-serial case, simulation is carried out for different wire lengths (1–8 mm) and in case of semi-serial, the simulation is performed for 4 mm long communication. All simulations are performed using 65 nm CMOS technology from STMicroelectronics with supply voltage of 1 V. Depending on the link circuits operating condition requirements, low-power low-vt or low-power high-vt transistors are used.

6.3.1 Wire Model and Simulation Waveforms

A distributed RLC-model is adopted to accurately model signaling over long on-chip wires. Furthermore, both capacitive and inductive coupling is added between wires to take into account crosstalk noise. The wire properties were set according to ITRS 65 nm technology node for global wiring [63]. In the serial data transmission wire modeling, wide and upper metal layers are assumed. Its wire width and separation distance were set to 1 μm and 1.5 μm, respectively. In the acknowledgment wire modeling, both wire width and separation distance were set to 210 nm. The RLC values of the wires were extracted using field solvers, FastHenry [45] and Linpar [46].

The simulation waveforms of major signals of the serial link are shown in Fig. 6.15. The two-phase bundled-data handshake signals of the two communicating

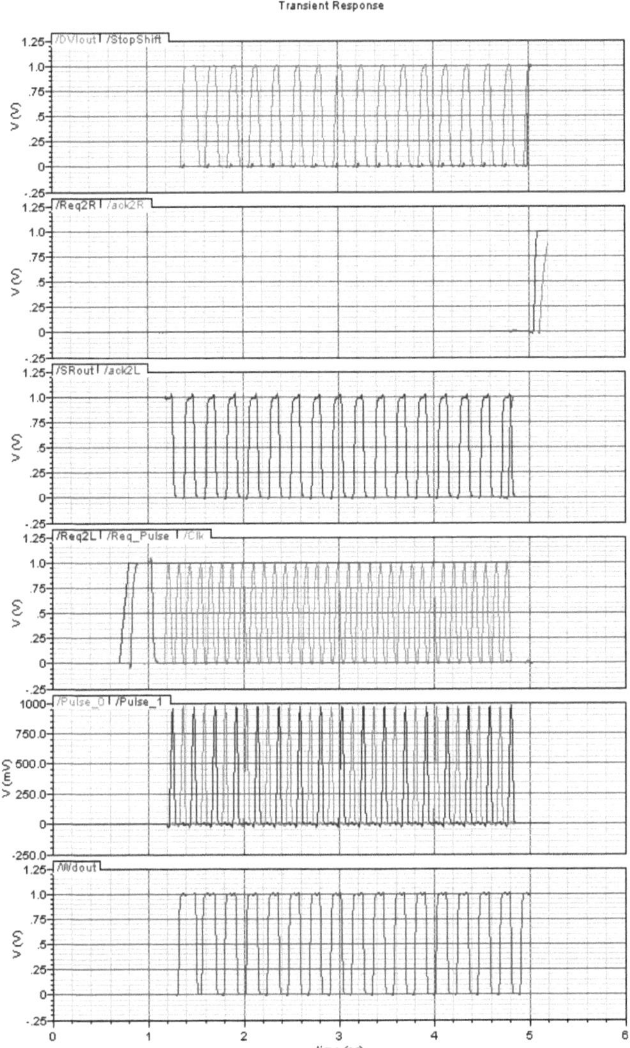

Fig. 6.15 Simulation waveforms of serial link

parties are *Req2L*, *Ack2L* and *Req2R* and *Ack2R* (see Fig. 6.4). The parallel load enable signal of the serializer's shift register is *Req_pulse*. The locally generated stoppable clock is *Clk* and the serialized data output of the serializer is *SRout*. The pulse dual-rail encoder outputs (the differential driver inputs) are *Pulse_1* and *Pulse_0*. *Wdout* and *DVIout* are the receiver and data validity decoder outputs, respectively. Signal which informs the receiving router the availability of all data for parallel output is *StopShift*.

6.3.2 Performance

The throughput of the presented bit-serial link is 9.091 Gbps (110 ps bit cycle) for all simulated link lengths and it is limited by the capacity of the clock generation circuit. It is known that the fastest clock that can be generated using a ring oscillator is bounded to $6 - 8\tau_4$ [85], resulting in 90 to 125 ps clock period for 65 nm technology. Bit shifting is performed at the negative edge of the clock and pulse dual-rail encoding during *high* state of the clock. The delay between consecutive data words is minimized since data loading to the shift register is performed while waiting for the acknowledgment to arrive from the receiver. As soon as the acknowledgment signal arrives, clock starts running and then shifting out the bit for transmission. The throughput of the semi-serial link increases linearly with the number of bit-serial parallel links. For instance, from eight parallel bit-serial links, a throughput of 72.728 Gbps was achieved. In the semi-serial link only one serializer control circuit is used for all serializers. Because of this, the locally generated clock is distributed through a clock tree. For a semi-serial link consisting of eight bit-serial links, the longest clock path is 23.5 μm. The RC values of the clock tree were extracted and the clock skew was so small in the simulations that it did not have effect on the link performance. On the other hand, the clock jitters had a direct impact in performance. Clock jitter and PVT variations are examined in Sect. 6.3.4.

6.3.3 Power and Energy Consumption

The bit-serial link is simulated for 32-bit word and the length of the link is varied from 1 to 8 mm. Its overall (including all link circuits) average power consumption and energy per bit is listed in Table 6.1. The power consumption of 32-bit serializer and deserializer is 2.198 mW and 1.416 mW, respectively. The power consumption of the link and its energy dissipation per bit do not increase steeply with the wire length because of the pulse signaling, where only a small portion of the wire needs to be charged. In addition, due to wider spacing of the wires the link has smaller coupling capacitance which allows the required driving current to be smaller and this in turn reduces the power consumed by the link.

It is known that if N number of the presented bit-serial links are used in parallel, the overall throughput of the channel becomes N times the throughput of one bit-serial link. But what about the power efficiency or energy dissipation per bit. In order to answer this question, a semi-serial link is designed and simulated. This link consists of eight bit-serial links each of which has 8-bit word length. The simulation results of the semi-serial link are then compared with 64-bit one bit-serial link. The wire length is varied from 1 to 8 mm for both semi-serial and bit-serial links simulations. As shown in Fig. 6.16, the semi-serial link has smaller energy dissipation than the bit-serial link. To be more precise, the energy dissipation of the semi-serial link is less than one-third of the bit-serial link. There are two reasons for this: first, proportional increase in channel bandwidth is much higher than the

Table 6.1 Power and energy dissipations of bit-serial communication

Link length [mm]	Average total power [mW]	Energy per bit [pJ/bit]
1	4.231	0.465
2	4.442	0.488
3	4.623	0.508
4	4.785	0.526
5	4.947	0.544
6	5.088	0.559
7	5.203	0.572
8	5.317	0.584

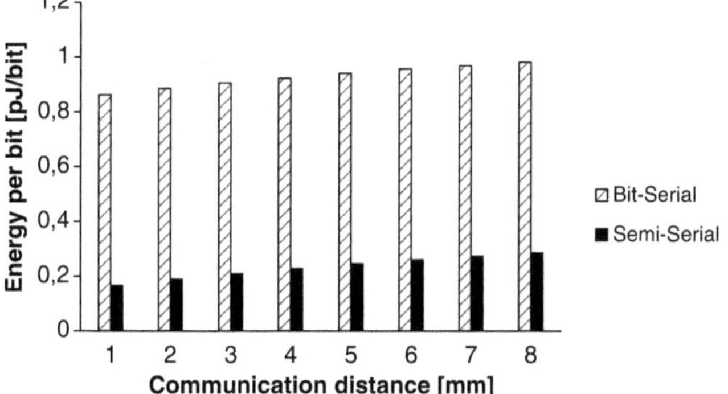

Fig. 6.16 Energy per bit of 64-bits word bit-serial and semi-serial links

increase in power consumption. Second, equal amount of power is dissipated in both links due to serializer/deserializer and their control circuits, because the size of the transmitted word is the same and there is no need to replicate the control and clock generating circuits. Only proper buffering is required since the locally generated clock is also used for pulse dual-rail data encoding. In addition, even if the semi-serial link has eight wires, the dynamic power consumption on the wire is reduced significantly due to the use of both pulse signaling and wave-pipelining. Therefore, semi-serial link is a better alternative for long-range high-throughput and energy efficient on-chip communication.

Assuming that the sending and receiving modules have 64-bit wide data and they are placed 4 mm apart each other, three different semi-serial links are simulated. The energy per bit of these links is presented in Table 6.2. While the throughput doubles and quadruples with the number of parallel links, the power consumption (energy) does not. The reason is that the control and clock generating circuits of the serializer are shared, and there is no need to replicate them for each link. This decreases the energy dissipation per bit of the semi-serial link as the number of parallel bit-serial links increases (Table 6.2). In case of eight parallel links, a throughput of 72.728 Gbps is achieved with 16.596 mW power consumption. If this

Table 6.2 Performance and energy dissipation of semi-serial links

Links	Throughput [Gbps]	Power consumption [mW]	Energy per bit [pJ/bit]
2	18.182	7.958	0.437
4	36.364	11.730	0.322
8	72.728	16.419	0.226

Table 6.3 Delay between *WDOUT* and *DVIOUT* signals

		Delay	
		t_{pHL}[ps]	t_{pLH}[ps]
Process corners	typ	51	51
	FF	35	33
	FS	44	42
	SS	52	80
Voltage	+10%	48	42
	−10%	76	33
Temperature	0	79	36
	100	86	43

result is compared with [83], which achieves a throughput of 67 Gbps with a power consumption of 150 mW for 16-bit word and 4 mm long bit-serial link, the power consumption of the proposed semi-serial link is almost one-tenth of [83] besides achieving a slightly higher throughput.

6.3.4 *Effect of PVT Variations*

Several simulations were carried out to assess the impact of PVT variations on the proposed link. The effect of the locally generated clock jitter on the sampling of the serializer and the encoding was examined by adding supply noise in the link circuit. Power supply noises of ±100 mV and ±150 mV with variable frequencies were used in the simulations. The worst-case clock jitters for 100 mV and –100 mV noises were –15 ps and 27 ps, respectively. In case of 150 mV and –150 mV, the jitters were –22 ps and 37 ps, respectively. In this case jitters, there was slight distortion in *SRdout* signal when it was high which caused the encoder pulse output to reach only up to 85% of VDD. However, the link still worked reliably up to ±200 mV noise.

Since the *DVIout* signal is used as a clock in the shift register of the deserializer, *Wdout* should be valid before *DVIout* and fulfill the setup time requirement. Data retrieval occurs faster than the data validity detection because in the data validity path there are current comparator and differential amplifier delays whereas in the data receiver only the differential amplifier delay (see Fig. 6.11). In order to ensure this timing constraint is met despite PVT variations, simulations were carried out for all process corners as well as by varying the supply VT. The delays between *Wdout* and *DVIout* signals are presented in Table 6.3. Though the amount of delay varied, the delay fulfilled the timing requirements for all cases.

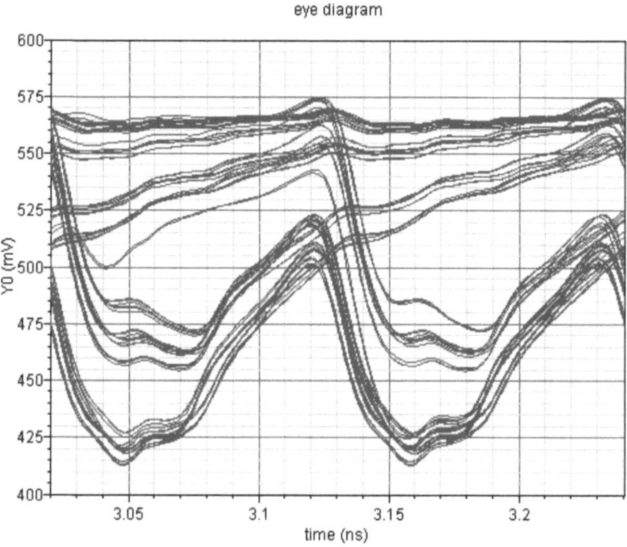

Fig. 6.17 Eye diagram of the receiver's differential input signals

6.3.5 Bit Error Rate (BER)

The BER of the presented link was estimated by performing 1000 Monte Carlo simulations in Spectre. In the simulations, pseudorandom bit stream of 213 bits were used as input data to the link. Also, power supply noise with amplitude of ±5%Vdd with variable frequencies, transient noise (thermal, shot, flicker), process variation and device mismatch were included in the simulation. The BER was determined by analyzing the eye opening size at the input of the receiver. A BER of 4.86E-12 was achieved when the output driver power consumption was 0.694 mW. When the output driver size was increased and it consumed 1.041 mW, the link error rate was decreased to 1.79E-13. The eye diagram of the receiver's differential input signals is shown in Fig. 6.17.

6.4 Fully Bit-Parallel vs Bit-Serial and Semi-Serial Links

With increasing number of complex, non-uniform sized nodes in a NoC, high-throughput low-power and area efficient long-range links become necessity. For example, the node size is 2 mm by 1.5 mm the same as in TERAFLOPS [69] and the NoC consists of 20 nodes. To connect the farthest nodes in regular mesh structure or for end-around channel of a torus an 8 mm long link is required. In order to analyze the trade-off between bit-serial, semi-serial and fully bit-parallel long-range channels of a NoC, three types of fully bit-parallel links were designed. Two of

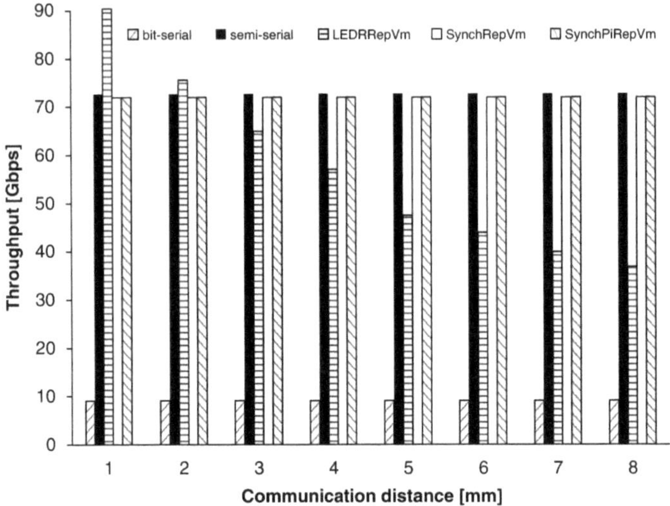

Fig. 6.18 Throughput of 64-bit word serial and parallel links

the links are synchronous while one is a self-timed delay-insensitive link. One of the synchronous fully bit-parallel links uses voltage-mode signaling with repeaters while the other one uses pipelining in addition to repeatered voltage-mode signaling. Also, a semi-serial link consisting of eight of the presented bit-serial links in parallel was designed and simulated. As the presented serial link is delay-insensitive, the self-timed bit-parallel link was also designed as a delay-insensitive link using two-phase LEDR encoding and optimally repeated voltage-mode signaling. The LEDR encoding was chosen due to its simpler and faster completion detection and data decoding logics than the conventional two-phase dual-rail encoding.

In the fully bit-parallel links, the wires were modeled narrower than the wires in the semi-serial links since they do not use high speed signaling which requires maximizing the inductance effect. The pitch of these link wires was two times the minimum pitch of a global wire according to ITRS [63]. In the self-timed LEDR encoded (*LEDRRepVm*) link, the required optimal number of repeaters and size of the repeaters were initially calculated using equation (36) of [53] and then the exact values were determined from simulations. The size of the optimal repeater was 48*minimum size inverter and it was inserted every 0.4 mm. In the two synchronous links, a clock cycle of 1.125 GHz was assumed so that a throughput of 72 Gb/s can be achieved from 64-bit parallel transmission, comparable to the semi-serial link. In the synchronous only repeatered voltage-mode (*SynchRepVm*) link, the repeaters were inserted every 0.4 mm and their size was determined so that it satisfied the clock cycle. In the pipeline and repeater based (*SynchPiRepVm*) link, pipeline stage and repeater were inserted every 2 mm and 0.4 mm, respectively.

To illustrate real life applications, all five links were designed to support 64-bit wide data transmission. As can be seen from Fig. 6.18, the throughput of

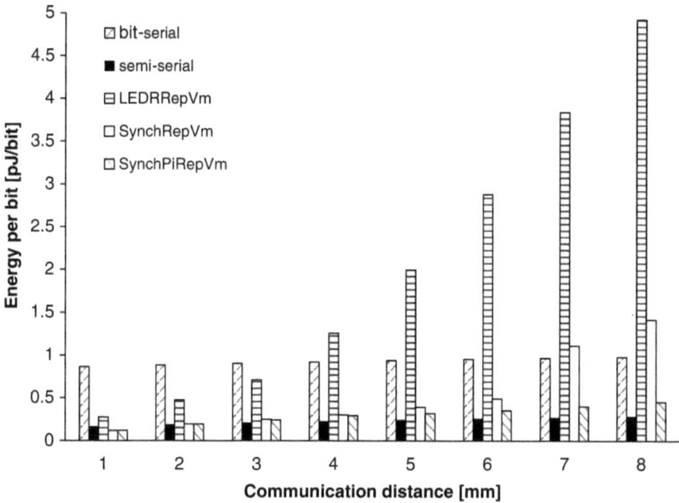

Fig. 6.19 Energy per bit of 64-bit word serial and parallel links

the *LEDRRepVm* link was greater than all the other links for 1 and 2 mm long communication distances. However, starting from 3 mm its throughput decreased rapidly with the increase in wire length, becoming lower than the throughput of the semi-serial and synchronous links. The reason for this is the need to transmit acknowledgment for each bit transmission. This makes the delay two times wire delay (forward and backward). Furthermore, the time required to carry out the completion detection of 64-bit transmission is significant though not dependent on the communication distance. If the throughput of the two self-timed delay-insensitive links is compared at 8 mm communication distance, the semi-serial link achieved 1.97 times the throughput of *LEDRRepVm*.

The energy dissipated per bit transmission for all the links is shown in Fig. 6.19. The semi-serial link dissipated the least energy starting from 3 mm long transmissions whereas *LEDRRepVm* link dissipated more energy than all the others for 4 mm and longer communications. The energy dissipation per bit of the *LEDRRepVm* link rose sharply with wire length, because the throughput decreased significantly while the power consumption increased with communication distance. It consumed 17.2 times more energy than the semi-serial link for 8 mm transmission. Among the fully bit-parallel links, *SynchPiRepVm* link dissipated the least energy per bit for all communication distances. This was due to requiring smaller sized repeaters in order to satisfy the clock frequency for the 2 mm pipelined segment. The percentage change of energy dissipations of bit-serial and fully bit-parallel links compared to the semi-serial link dissipation are presented in Table 6.4. At 4 and 8 mm transmissions, the *SynchPiRepVm* link, which is the lowest among fully bit-parallel links, dissipated 29% and 59% more energy than the semi-serial link.

Table 6.4 Percentage change of energy per bit dissipation compared to semi-serial link

Wire length [mm]	Bit-serial [%]	*LEDRRepVm* [%]	*SynchRepVm* [%]	*SynchPiRepVm* [%]
1	417	65	−28	−28
2	366	151	3	3
3	331	239	20	15
4	305	453	33	29
5	283	712	60	30
6	267	1005	89	36
7	254	1302	307	47
8	243	1622	396	59

Table 6.5 Area comparison between serial and parallel 64-bit links

Link type	Active area [μm^2]	Wiring area [μm^2]
Bit-Serial	84	33040
Semi-Serial	91	313040
LEDRRepVm	6854	428400
SynchRepVm	592	215040
SynchPiRepVm	527	215040

The required active and wiring areas of the five 64-bit word links for 8 mm long transmission are shown in Table 6.5. The active area taken by the *LEDRRepVm* link is much larger than the others. It takes about a 75 times larger active area than the *semi-serial* link. The reason is that the LEDR encoder along with repeaters uses 98% of its total active area. The one-bit LEDR encoder consists of three double-edge triggered flip-flops, five inverters and one XOR gate. The synchronous bit-parallel links also take a larger area than the semi-serial and bit-serial links. Their area is about a six times larger than the area of the semi-serial link. The wiring area of the *LEDRRepVm* link is also larger than the other links because it needs $2N$ wires for N-bit transmission. The semi-serial link takes only 73% of the wiring area compared to the *LEDRRepVm* link. However, it requires 46% more wiring area than the synchronous links. This is due to both the use of wider wires and differential signaling. The synchronous bit-parallel links are single-ended and does not support delay variation insensitive data transmission.

6.5 Chapter Summary

In this chapter design and analysis of a high-throughput and low-power serial on-chip communication link is presented. The developed link has used novel design techniques, circuits and architecture which have enabled the link to achieve high-throughput with low energy dissipation. The link is self-timed and designed using high-speed serialization/deserializtion and pulse dual-rail encoding techniques.

Wave-pipelined differential pulse current-mode signaling was also employed in order to maintain the throughput with low power consumption. This link is a promising candidate for long-range NoC channels, which are needed inherently due to topologies or through customization of regular 2D networks. In addition, its delay-insensitive data transfer makes it more appropriate for nanoscale NoC interconnects where delay variations are inevitable. Our simulation results showed that the semi-serial link outperformed the fully bit-parallel and bit-serial links in energy efficiency starting from 2 mm communication distances. In 64-bit 8 mm data transmission, the semi-serial link achieved a throughput of 72.72 Gbps with 286 fJ/bit energy dissipation, whereas the fully bit-parallel synchronous link with the same throughput dissipated five times more energy. The PVT variation analysis showed that the proposed link worked reliably for all process corners, $\pm 10\%$ supply voltage variations and 0 to 100^{deg}C temperature range. The proposed semi-serial link also needed the smallest active area compared to any of the considered bit-parallel links and a smaller wiring area than the self-timed bit-parallel link.

Chapter 7
Comparison of the Designed Interconnects

In the previous three Chaps. 4, 5, and 6, design and analysis of delay-insensitive and high-performance on-chip interconnects have been presented. These interconnects are suitable for any kind of point-to-point on-chip communication, such as in a SoC to connect nearby or far away system blocks and in a NoC between two routers. The purpose of this chapter is to make a generalized summary of the presented interconnects as well as comparisons between them. In order to do so, all interconnects are redesigned and simulated in 65 nm CMOS technology from STMicroelectronics with 1V supply voltage.

7.1 Summary of the Interconnects

The presented four interconnects use different encoding/decoding, completion detection and signaling techniques. Each approach has its own advantages and limitations but all have the same goals: delay variation robustness, high performance and energy efficiency. The generalized summary of these interconnects is presented in Table 7.1. *LEDRCm*, *PMCmFCD*, and *DualdiffFCD* interconnects are designed for fully bit-parallel transmission. Whereas the *Bit-Serial* interconnect is for serial transmission of bits and the *Semi-Serial* is made up of few bit-serial interconnects in parallel. In *Bit-Serial* and *Semi-Serial* interconnects the bits are wave-pipelined on the wire and acknowledgment is transmitted per word.

7.2 Comparison of the Interconnects

In this section, comparison between *LEDRCm*, *PMCmFCD*, *DualdiffFCD*, *Bit-Serial*, and *Semi-Serial* interconnects in terms of performance, energy and area will be carried out for a number of transmissions with different bit widths. Except in the

E.E. Nigussie, *Variation Tolerant On-Chip Interconnects*, Analog Circuits and Signal Processing, DOI 10.1007/978-1-4614-0131-5_7,

Table 7.1 Summary of interconnects

Interconnect	Encoding	Signaling	Advantages	Limitations
LEDRCm	LEDR	Binary current sensing	No resetting transitions. Level based detection rather than transition. Better performance and energy efficiency due to current sensing signaling.	It is not possible to use the proposed fast completion detection scheme and to shut wire currents off in order to save power
PMCmFCD	Two-phase 1-of-4	Pulsed Multilevel current sensing	No resetting transitions. Performance enhancement due to current sensing signaling and fast completion detection. Power saving due to shutting off wire currents.	Extra design effort is required to determine the optimal current margins
DualdiffFCD	Two-phase dual-rail	Multilevel and differential current sensing	No resetting transitions. Area and power efficient integration of dual-rail encoding and differential signaling. High common-mode noise rejection. Performance and energy efficiency improvement due to differential current sensing signaling and fast completion detection.	Extra design effort is required to determine the optimal current margins
Bit-Serial Semi-Serial	Pulse dual-rail	Wave-pipelining, pulse signaling, and differential current mode signaling	Fast encoding. No data decoding logic. Per word acknowledgment without losing bit-level delay-insensitivity. High common-mode noise rejection. High throughput and energy efficient due to integration of signaling techniques as well as circuit optimizations.	Not energy efficient for short-range links

Table 7.2 Throughput of interconnects

	Throughput [Gbps]		
Communication distance [mm]	*LEDRCm*	*PMCmFCD*	*DualdiffFCD*
1	5.988	6.92	7.936
2	4.662	4.784	5.714
3	3.809	3.861	4.705
4	3.205	3.262	3.968
5	2.597	2.828	3.322

case of the serial link, acknowledgment signal transmission is performed using the same signaling scheme and circuits (the same as the one presented in Sect. 5.3.3). In the serial communication, acknowledgment is sent per word using differential pulse current-mode signaling. In order to have a proper comparison between the bit-parallel interconnects, their drivers are designed so that they have the same output current values. Since the *LEDRCm* interconnect uses binary signaling while the other two use multilevel current sensing signaling, the driver output current of *LEDRCm* is set to one of the current levels (I_2) of the other two interconnects.

7.2.1 Performance

In the *LEDRCm* interconnect the proposed high-speed completion detection technique is not implemented, because data validity decoding can be done only by detecting transitions, as LEDR encoding is based on data phase and state. The *PMCmFCD* and *DualdiffFCD* interconnects use the high-speed current-mode completion detection circuit. The throughputs of the three interconnects are listed in Table 7.2 for 2-bit transmission and 1–5 mm long communication distances. Two-bit transmission is considered in order to have a proper comparison with the 1-of-4 encoded link, *PMCmFCD*, since the smallest possible transmission in the 1-of-4 encoded link is two bits. When the communication distance increases the throughput of all the interconnects decreases. The *DualdiffFCD* interconnect achieves the highest throughput among the three interconnects because of its differential signaling scheme. The throughput of *PMCmFCD* interconnect is higher than that of the *LEDRCm* link. The throughput of bit-serial and semi-serial links is not affected by the wire length as they use wave-pipelining, and an acknowledgment is sent per word instead of per bit. The *Bit-Serial* link achieves a throughput of 9.09 Gbps and the throughput of *Semi-Serial* link increases linearly with the number of parallel bit-serial links.

The current sensing interconnects have also been designed and simulated for a number of different bit widths from 2 to 64 bits. The communication distance is assumed as 2 mm for all the interconnects. Their throughput is shown in Fig. 7.1. For 2- and 4-bit transmissions there is no big difference between the throughputs. Starting from 8-bit transmission the current sensing interconnect's throughput

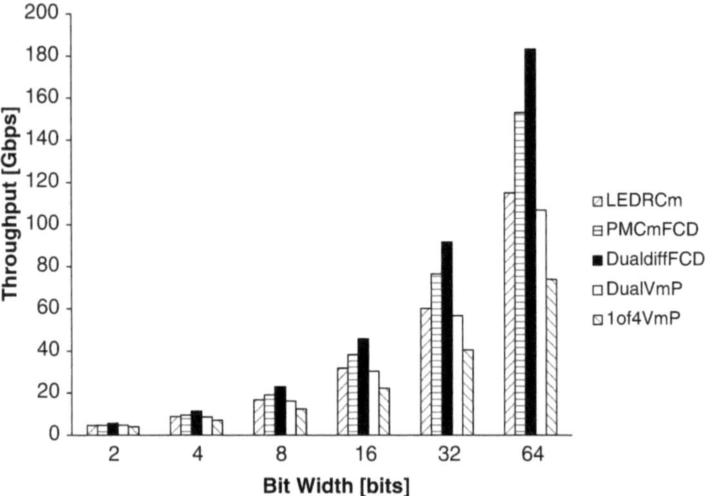

Fig. 7.1 Throughput versus bit width of links

Table 7.3 Throughput of 64-bit word 5 mm long links

Link name	Throughput [Gbps]
PMCmFCD	90.523
DualdiffFCD	106.312
Bit-Serial	9.09
Semi-Serial8	72.72
Semi-Serial12	109.08
Semi-Serial16	145.44

becomes higher than the pipelined voltage-mode links. The throughput gap between the current sensing and pipelined voltage-mode links increases with the bit width. For example, the throughput of *DualdiffFCD* is 1.4 and 1.7 times the throughput of *DualVmP* for 8- and 64-bit transmissions, respectively. The *DualdiffFCD* link achieves the highest throughput because of the differential signaling, followed by the *PMCmFCD* link. In 64-bit transmission the throughput of *DualdiffFCD* is 1.19 and 1.59 times *PMCmFCD* and *LEDRCm* links throughput, respectively. For 32- and 64-bit transmissions the *LEDRCm* link's throughput is considerably lower than that of the two current sensing links due to the difference in completion detection.

The purpose of the serial link is to communicate blocks of data over a long distance. Due to this, the comparison with the bit-parallel current sensing links is carried out for 64-bit 5 mm long transmission. Among the serial communication links one bit-serial and three semi-serial links are considered. The *Semi-Serial8*, *Semi-Serial12*, and *Semi-Serial16* links in Table 7.3 are links consisting of 8, 12, and 16 bit-serial links in parallel, respectively. The throughputs of these interconnects are presented in Table 7.3. Compared to the *Bit-Serial* and *Semi-Serial8* links, the two bit-parallel links perform better.

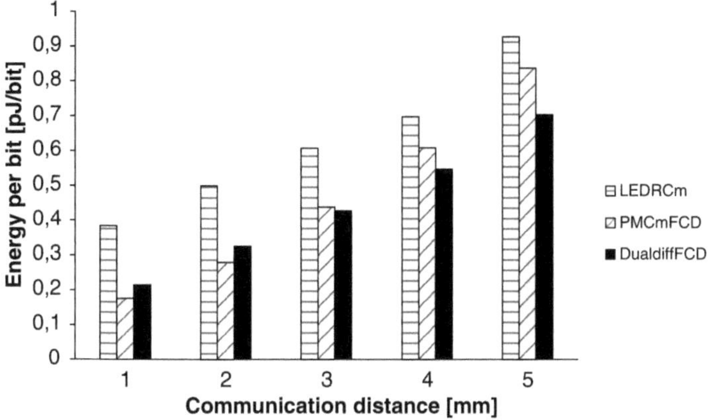

Fig. 7.2 Energy per bit versus communication distance of interconnects

7.2.2 *Power Efficiency*

The energy per bit dissipation of *LEDRCm*, *PMCmFCD*, and *DualdiffFCD* interconnects are determined for 1 to 5 mm long transmissions. As it can be seen in Fig. 7.2, the *LEDRCm* interconnect dissipates the highest amount of energy per bit for all communication distances. One of the reasons for this is that it uses voltage-mode completion detection which contains a large number of power-consuming logic gates. The *PMCmFCD* interconnect dissipates lowest energy per bit at 1 and 2 mm long communications and after that its energy consumption becomes higher than that of the *DualdiffFCD* interconnect. This is due to the single-ended signaling used in the *PMCmFCD* interconnect.

As the purpose of these links is to provide transfer of data between two points at a global distance with minimum possible energy, their energy per bit consumption is analyzed for 2 to 64-bit 2 mm long transmission. Besides the three current sensing links, the conventional two-phase dual-rail and 1-of-4 encoded and optimally pipelined voltage-mode interconnects' energy dissipation per bit is examined and presented. The *LEDRCm* interconnect dissipates the highest energy per bit and the *DualdiffFCD* link dissipates the least energy starting from 4-bit transmission (Fig. 7.3). For instance, 64-bit transmission using *LEDRCm* interconnect dissipates 0.507 pJ/bit which is 2.96 times higher than the energy per bit dissipation of *DualdiffFCD* link. For 2-bit transmission the pipelined voltage-mode interconnects are preferable, consuming the lowest energy. Starting from 8-bit transmission both *PMCmFCD* and *DualdiffFCD* links consume the lowest energy, thanks to the current-mode completion detection scheme.

The *Bit-Serial* and *Semi-Serial8* links dissipate 0.886 pJ/bit and 0.190 pJ/bit energy, respectively for 64-bit 2 mm long transmission. The semi-serial link dissipates almost the same energy as the *PMCmFCD* link and 11.1% more energy compared to the *DualdiffFCD* link.

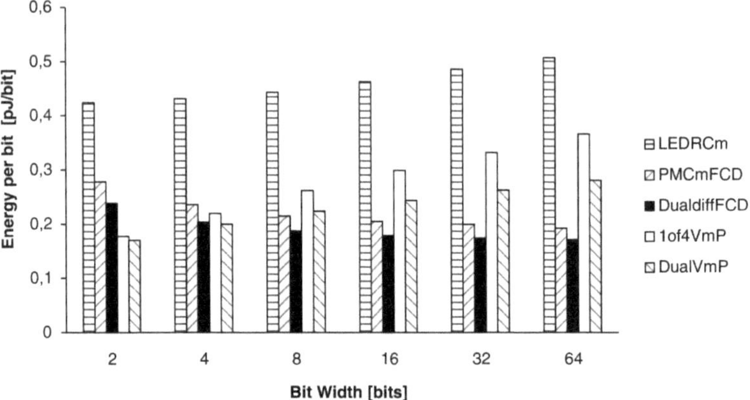

Fig. 7.3 Energy per bit versus transmission bit width of links

Table 7.4 Area comparison of 64-bit 2 mm long transmission links

Link name	Silicon area [μm^2]	Wiring area [μm^2]
LEDRCm	1714	55020
PMCmFCD	508	55020
DualdiffFCD	599	55020
1of4VmP	704	66780
DualVmP	662	66780
Bit-Serial	84	8260
Semi-Serial8	91	78260

7.2.3 Area

In today's interconnect-centric and ultra integration era, area is among the fore front design parameters. Comparison of silicon and wiring area is performed for the *LEDRCm*, *PMCmFCD*, *DualdiffFCD*, *Bit-Serial* and *Semi-Serial8* interconnects for 64-bit 2 mm long communication. The active area taken by the *LEDRCm* is 237% and 186% more than the area of *PMCmFCD* and *DualdiffFCD* interconnects. The reason is that one bit LEDR encoding requires three double-edge triggered flip flops and one 2-input XOR gate. In addition, 63 2-input C-elements are needed for the completion detection. The active area required by *PMCmFCD* and *DualdiffFCD* links is also smaller than the pipelined voltage-mode links but larger than the serial links. The *LEDRCm*, *PMCmFCD* and *DualdiffFCD* links take only 82% and 58% wiring area of the pipelined voltage-mode and *Semi-Serial8* links, respectively (Table 7.4).

7.3 Chapter Summary

Generalized summary of the three bit-parallel and serial interconnects that were presented in the previous three chapters has been discussed. Comparisons of performance, energy per bit and area have also been carried out. Among the bit-parallel interconnects dual-rail encoded differential current sensing interconnect achieves the highest throughput with lowest energy per bit dissipation. The LEDR encoded current sensing interconnect has the poorest performance and highest energy per bit dissipation. The dual-rail differential current sensing interconnect takes slightly larger active area. The semi-serial link outperforms the bit-parallel ones starting from 5 mm communication distance in terms of throughput, showing its potential for long-range communication.

Chapter 8
Circuit Techniques for PVT Variation Tolerance

As part of an IC, on-chip interconnects experience two types of variations: physical and environmental. A physical variation is due to the manufacturing process imperfections. Whereas, environmental variations occur during the operation of a circuit and includes dynamic variations in the supply VT. Precise control of the manufacturing process is worsening with technology scaling due to smaller dimensions, smaller number of doping atoms and aggressive lithographic techniques. This becomes a major concern since it causes uncertainty in electrical characteristics of devices and interconnecting wires which consequently affect the reliability of the system. Variability in the operating environment also affects the reliability of on-chip interconnects. As the variations increase, techniques which reduce their impacts while providing the highest performance for a given power constraint are necessary at the system, architecture, and circuit levels [116]. In this chapter, circuit level techniques which ensure signal integrity of a current sensing on-chip interconnect in the presence of PVT variations are developed and implemented. Since all the interconnects that are presented are delay-insensitive, the developed signal integrity technique considers only the signal amplitude variation.

This chapter is organized as follows. In the next section, signal integrity problems of a current sensing interconnect due to process and environmental variations are discussed. Brief discussion about post-manufacture variation adaptation technique is presented in Sect. 8.2. Process variation tolerance technique along with its algorithm, methodology, and circuit realization is presented in Sect. 8.3. The runtime environmental variation monitoring and management technique is discussed along with its implementation in Sect. 8.4. Simulation results of the presented PVT variation tolerance techniques as well as analysis of power, delay and area overheads are presented in Sect. 8.5. The summary of the chapter is presented in the last section.

E.E. Nigussie, *Variation Tolerant On-Chip Interconnects*, Analog Circuits and Signal Processing, DOI 10.1007/978-1-4614-0131-5_8,

8.1 Signal Integrity of Current Sensing Interconnect

For convenience a current sensing interconnect can be divided into three parts: driver, wire, and receiver. Figure 8.1 shows a current sensing interconnect structure along with its electrical parameters which can be affected by PVT variations. In a current sensing interconnect the receiver compares wire current with a reference in order to retrieve the data transmitted from the far-end. If the variation of input and/or reference current is out of the allocated margin then it can lead to erroneous output. It is possible to allocate large current margins by considering worst-case variations, however; this has power consumption costs associated, especially with the increase in the number of variability sources. Thus, it is wise to deal with variations in these two currents and devise techniques at the circuit level, which can tolerate their process and environmental induced variations thereby enhancing the reliability of the interconnect with low power overhead.

8.1.1 *Effects of Process Variation*

Both wire and reference currents may deviate from their nominal values due to uncertainties in front-end and back-end manufacturing processes. The front-end process comprises of manufacturing steps that are involved in creating devices, while the back-end is responsible for creating the interconnecting wires between the devices. The primary causes of variations in device electrical parameters are threshold voltage(V_{th}) variation, line-edge roughness (channel length and width variations), oxide thickness variation and dopant fluctuations [119–121]. These variations make the output current of the driver I_{win} to be different from its nominal value (Fig. 8.1).

It is known that sub-100 nm CMOS transistors are velocity saturated, i.e., there is a linear dependence between I_D and V_{GS} in the strong-inversion region [131]. Also, threshold voltage is strongly impacted by channel length and operational voltage V_{DS}. The output current of the driver I_{win} under velocity saturation can be expressed by the Equation 8.1 [131, 132]. E_C is the critical electric field at which the carrier velocity becomes saturated. From this equation it can be seen that variations in

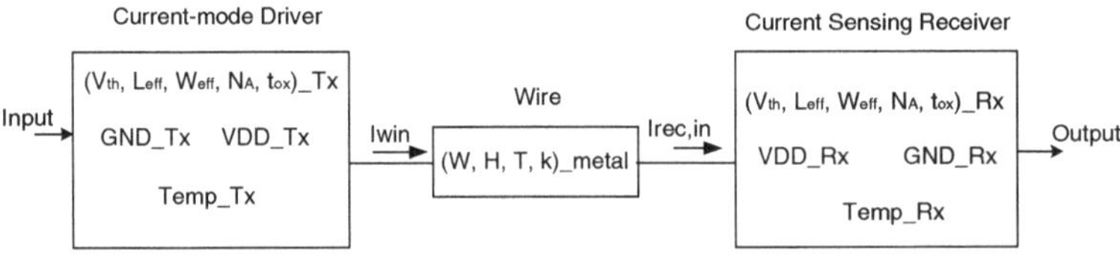

Fig. 8.1 Variable parameters of current sensing interconnect

threshold voltage, channel length and width and/or oxide thickness make I_{win} to fluctuate from its intended value.

$$I_{DSat} = \frac{W}{L}\frac{\mu_{eff}C_{ox}}{2}V_{DSat}(V_{GS} - V_{TH}) \tag{8.1}$$

$$where\ V_{DSat} = \frac{(V_{GS} - V_{TH})E_C L}{(V_{GS} - V_{TH}) + E_C L} \quad and\ C_{ox} = \frac{\epsilon_{ox}}{t_{ox}}$$

In order to examine the effect of front-end process variations on signal integrity of *LEDRCm*, *PMCmFCD*, and *DualdiffFCD* interconnects, Monte Carlo process runs of 1000 are carried out using 65 nm technology statistical model from STMicroelectronics. The communication distance is assumed to be 2 mm long. The variation of output current of (*LEDRCm*) interconnect driver is shown in Fig. 8.2a. In this simulation, only the front-end variability in the data encoder and the driver is considered. Its worst-case variation from its mean is 20 μA. The effect of electrical parameter variations of the termination transistor on I_{win} is also simulated and shown in Fig. 8.2b. Variability in the termination load causes additional I_{win} variation, taking the total I_{win} variation to 40 μA. The variation in the receiver's input current $I_{rec,in}$ due to fluctuations in the data encoder, driver and termination transistors is shown in Fig. 8.2c. Its worst-case variation is about 50 μA, and hence requiring a current margin greater than that, in between $I_{rec,in}$ and receiver's reference currents.

The variation of *PMCmFCD* I_{win} considering the process variations of the data encoder and the driver devices is shown in Fig. 8.3a. In this case its I_{win} worst-case variation is 40 μA. Variations in its I_{win} and $I_{rec,in}$ when the termination transistor effect is included are shown in Figs. 8.3b and 8.3c, respectively. Worst-case variation of I_{win} is 40 μA, while $I_{rec,in}$ is about 50 μA.

The *DualdiffFCD* interconnect's I_{win} and $I_{rec,in}$ variations due to front-end variabilities were also examined. Its I_{win} variation due to transmitter side device parameters uncertainties is shown in Fig. 8.4a and its worst-case variation was about 50 μA. Variations of I_{win} and $I_{rec,in}$ due to manufacturing variabilities of encoder, driver and termination transistors are shown in Figs. 8.4b and 8.4c, respectively. I_{win} variation increases by 25 μA due to the termination transistor. The current variations in this interconnect were larger than those in *PMCmFCD* and *LEDRCm* interconnects.

The fluctuations in the back-end processes cause variations in geometry and material properties of the wire structure. Studies show that among the back-end process steps, erosion and dishing during CMP process has strong impact on wire parasitics. This is due to the systematic pattern or spatial effects (metal density, width and space) [124]. In general, dishing strongly affects wide lines, while erosion is worse for narrower oxide and dielectric spacing between lines. In medium size features, the two effects combine, so that both dishing and erosion contribute to overall copper thickness reduction. The strong correlation between metal width and

Fig. 8.2 *LEDRCm* interconnect I_{win} and $I_{rec,in}$ variations. (**a**) I_{win} variation due to encoder, driver and termination device parameters variability. (**b**) I_{win} variation due to encoder, driver and termination device parameters variability. (**c**) $I_{rec,in}$ variation due to encoder, driver and termination device parameters variability

a

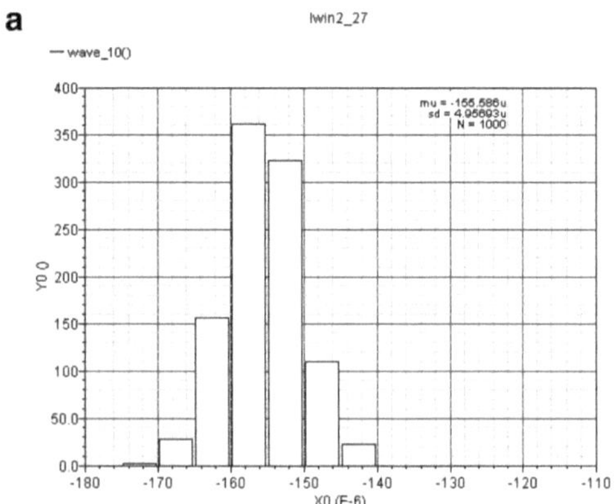

I_{win} variation due to encoder and driver device parameters variability.

b

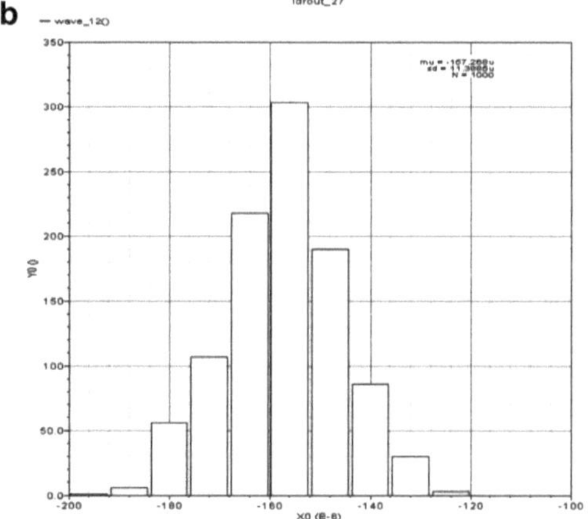

I_{win} variation due to encoder, driver and termination device parameters variability.

c

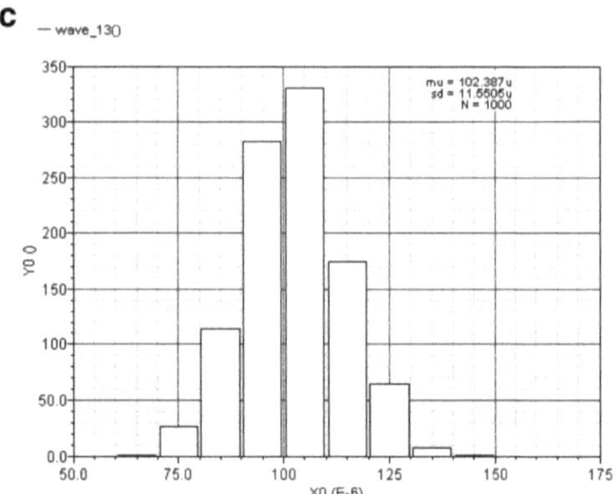

$I_{rec,in}$ variation due to encoder, driver and termination device parameters variability.

Fig. 8.3 *PMCmFCD* interconnect I_{win} and $I_{rec,in}$ variations. (**a**) I_{win} variation due to encoder and driver device parameters variability. (**b**) I_{win} variation due to encoder, driver and termination device parameters variability. (**c**) $I_{rec,in}$ variation due to encoder, driver and termination device parameters variability

a

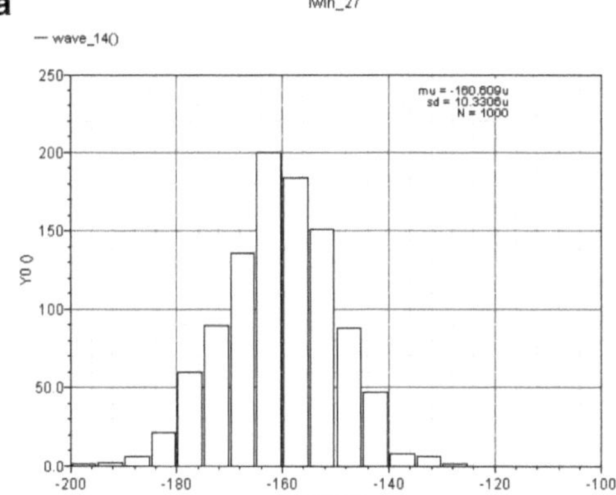

I_{win} variation due to encoder and driver device parameters variability.

b

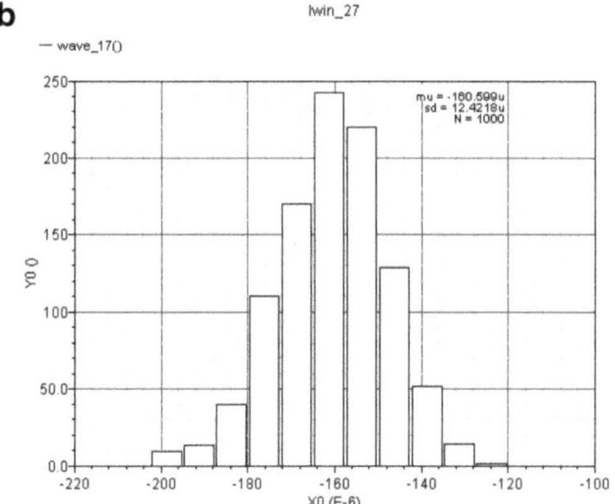

I_{win} variation due to encoder, driver and termination device parameters variability.

c

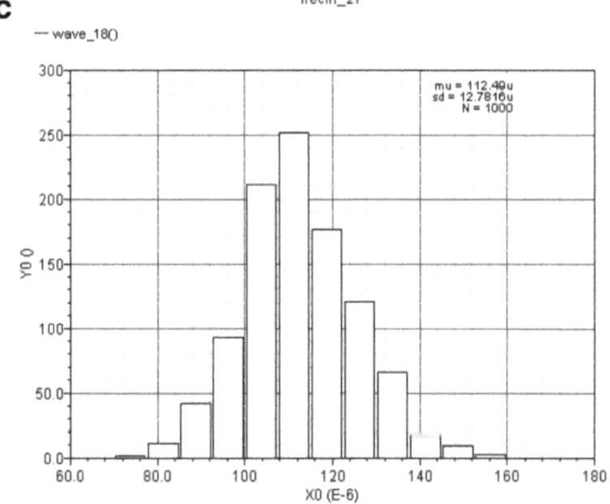

$I_{rec,in}$ variation due to encoder, driver and termination device parameters variability.

Fig. 8.4 *DualdiffFCD* interconnect I_{win} and $I_{rec,in}$ variations. (**a**) I_{win} variation due to encoder and driver device parameters variability. (**b**) I_{win} variation due to encoder, driver and termination device parameters variability. (**c**) $I_{rec,in}$ variation due to encoder, driver and termination device parameters variability

a

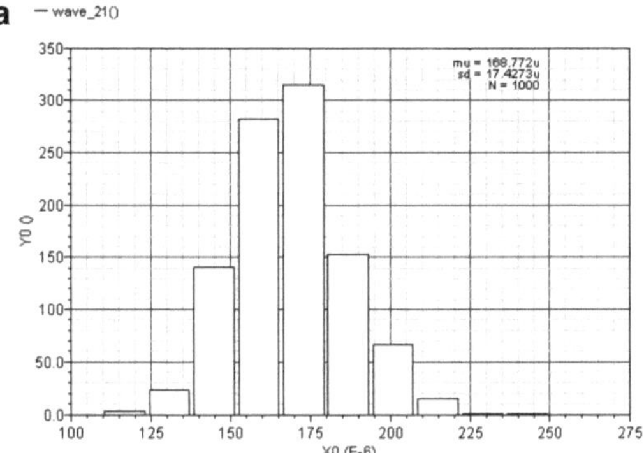

I_{win} variation due to encoder and driver device parameters variability.

b

Iwin_27

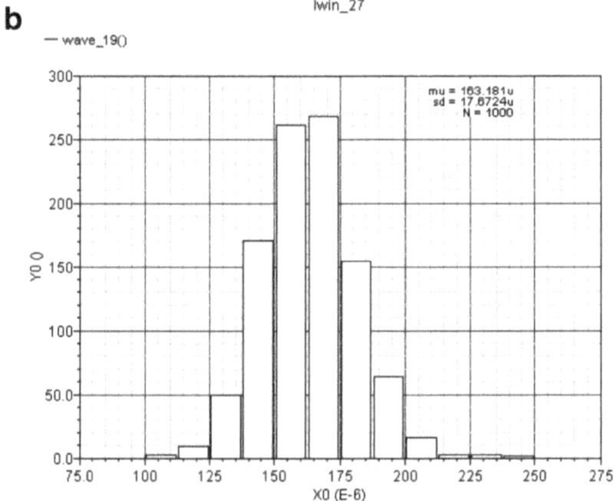

I_{win} variation due to encoder, driver and termination device parameters variability.

c

Irecin_27

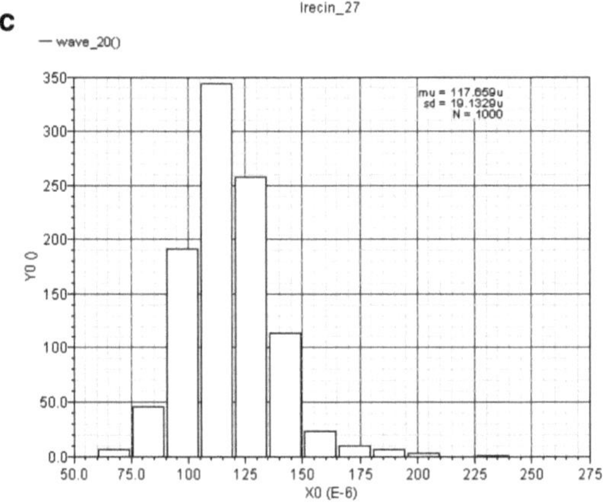

$I_{rec,in}$ variation due to encoder, driver and termination device parameters variability.

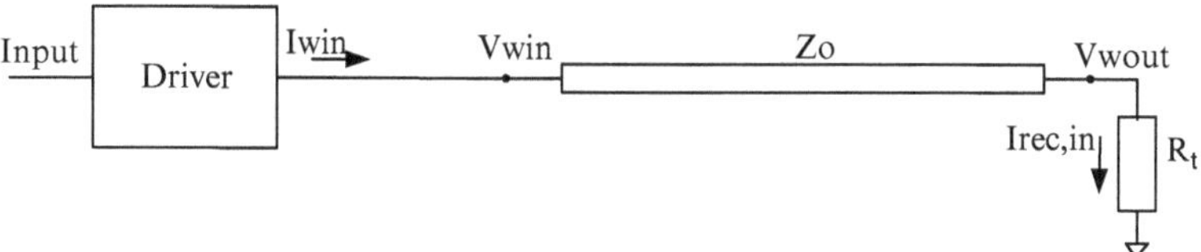

Fig. 8.5 Interconnect model for analysis

thickness variation due to CMP has also been proved from test chip measurements in [123]. The effect of line loss from dishing and erosion can be considerable and directly impacts the resulting electrical parameters of interconnecting wires. In [122], increase in resistance was observed on wide lines due to dishing and on high pattern densities due to dielectric erosion. Parasitic resistance, capacitance and inductance of the wire vary because of variations of metal and inter-layer dielectric (ILD) thickness and width as well as due to variations of material properties such as resistivity. It has been shown, that a 10% increase in width leads to about 10% increase in total capacitance, 12% increase in coupling capacitance, and 10% reduction in resistance [130]. It has been demonstrated, that parasitic RLC variations affect circuit performance [125–129].

In a current sensing interconnect, wire parasitics variation may impact signal integrity by causing variation in driver's output current and receiver's input current. Driver's output current I_{win}, is affected by the variation in its load (effective impedance which includes wire parasitics and termination load). The receiver's input current $I_{rec,in}$, is usually different from the driver's output current due to the non-ideal behavior of the interconnecting wire. These parameters are different from the ones estimated at design time due to variation of wire parasitics. The simple model of a current sensing interconnect demonstrating the variation in $I_{rec,in}$ is shown in Fig. 8.5. Let us assume that the characteristic impedance of the lossy transmission line is Z_o and the current through the line is I_{win}. The near and far-end voltages are V_{win} and V_{wout}, respectively where V_{wout} and $I_{rec,in}$ can be expressed as follows:

$$V_{wout} = V_{win} - I_{win} Z_o \tag{8.2}$$

$$I_{rec,in} = \frac{V_{wout}}{R_t} \tag{8.3}$$

From these two equations, it can be seen that $I_{rec,in}$ can be different from its nominal value determined at design time, because of variations of wire parasitics (Z_o) and termination device parameters. It has already been demonstrated from the Monte Carlo simulations that variations of termination transistor parameters cause additional variations in $I_{rec,in}$. The receiver's reference current also deviates from its nominal value due to variation of its devices electrical parameters. The conventional approach to ensure the reliability of the interconnect is taking into account effects of all these variations at design time and allocating large enough current margin. However, this has considerable power consumption cost, this is especially significant for multilevel current sensing interconnects. Hence, one has to further explore alternative power efficient technique.

8.1.2 Runtime Supply Voltage and Temperature Variations

The impact of delivering increasing currents to the huge number of active devices on a chip, and the effect of parasitics on both on-chip and package power delivery wires, leads to deviation of V_{DD} and GND signals from their nominal values. Increasing operating frequencies and power densities in sub-100 nm high performance ICs leads to an increase in voltage drops in the power grid. For instance, a voltage drop of 18% of the nominal voltage has been reported in $POWER6^{TM}$ dual-core processor fabricated in 65 nm SOI processes [133]. In the multicore scenario, clock-gating, power-gating and other power saving techniques have undesired consequences like increase in the variations of current drawn by different cores leading to additional supply voltage fluctuations. The semiconductor industry has already moved to dynamic voltage drop (DVD) analysis [134] in order to account for the contribution of power density, variations in switching activity profile and impact of inductance and decaps. DVD also captures the impact of spatial and temporal switching events. This move shows the importance of taking into account the unavoidable temporal and spatial voltage drop fluctuations which may lead to signal integrity problems for on-chip communications if they are not addressed well. Moreover, temperature variability rises due to the distributed nature of an IC, and due to the fact that some components dissipate more power than others. In the silicon substrate, heat generated at one point spreads and causes an increase in temperature at nearby points. Temperature variations also occur with time, as the subsystems switch between idle and active periods. While designing the current sensing interconnects, temperature variations must be considered as they affect the device and wire characteristics.

Runtime supply VT variations cause fluctuation in device's drain current. To characterize the drain current fluctuations induced by environmental variations, BSIM4 MOSFET current equations are used [135].

$$I_{ds} \propto \frac{I_{ds0}}{1 + R_{ds} I_{ds0} / V_{dseff}} \tag{8.4}$$

$$I_{ds0} \propto \frac{V_{gsteff} \mu_{eff} V_{dseff} (1 - A_{bulk} V_{dseff} / 2(V_{gsteff} + 2V_T))}{1 + V_{dseff} / E_{SAT} L_{eff}} \tag{8.5}$$

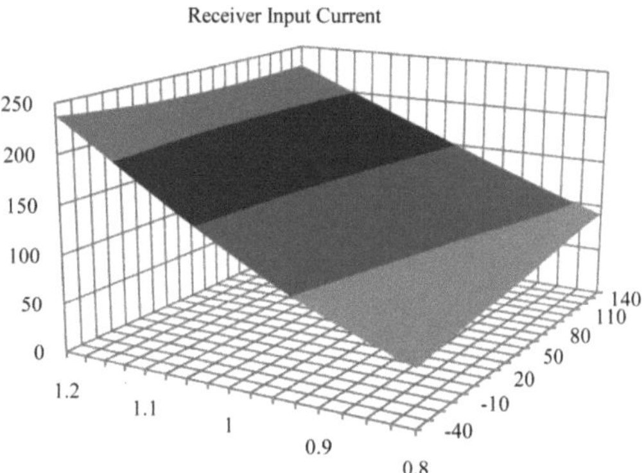

Fig. 8.6 $I_{rec,in}$ versus supply voltage and temperature

Where I_{ds}, I_{ds0}, R_{ds}, V_{dseff}, V_{gsteff}, A_{bulk}, μ_{eff}, V_T, E_{SAT}, and L_{eff} are the drain current with short-channel effects, drain current of a long channel device, parasitic drain-to-source resistance, effective drain-to-source voltage, effective gate overdrive ($V_{GS} - V_t$), parameter to model the bulk charge effect, effective carrier mobility, thermal voltage, electric field at which the carrier drift velocity saturates and effective channel length, respectively. MOSFET channel current is a function of both gate and drain voltages. Either of these voltages, or both, are affected by supply voltage variations depending on the circuit configuration. These variations in turn affect the drain current. Absolute values of threshold voltage, carrier mobility, and saturation velocity degrade with the increase in temperature [135, 136]. The degradation of threshold voltage with temperature tends to increase the drain current due to the increase in gate overdrive ($V_{GS} - V_t$), whereas degradation in carrier mobility tends to reduce the drain current as can be seen from Equation 8.5. Hence, overall variation of I_{win} is determined from cumulative variation of V_{GS} and V_{DS} caused by supply voltage fluctuation and the variation of the dominant device parameter when the temperature varies. Furthermore, the resistivity of a wire increases with temperature [137], increasing the parasitic resistance of the interconnecting wires which in turn decreases $I_{rec,in}$. It is also affected by the supply VT variations at the transmitter end (Fig. 8.1).

As an example, the effect of VT variations on *PMCmFCD* interconnect's $I_{rec,in}$ and reference current have been examined in Cadence Analog Spectre using 65 nm CMOS technology from ST Microelectronics. The VT was swept by 25 mV and 10^{deg}C, respectively. In $I_{rec,in}$ analysis, supply VT at the transmitter end were varied, while in reference current analysis, VT at the receiver end were varied. Both analyses show that voltage fluctuation has much more pronounced effect on current variation than temperature (Figs. 8.6 and 8.7). For instance, around the nominal operating point a ± 100 mV change in supply voltage at the transmitter end causes about $\pm 40\mu$A variation in $I_{rec,in}$ whereas a temperature increase of 100^{deg}C causes only about $\pm 13\mu$A variation.

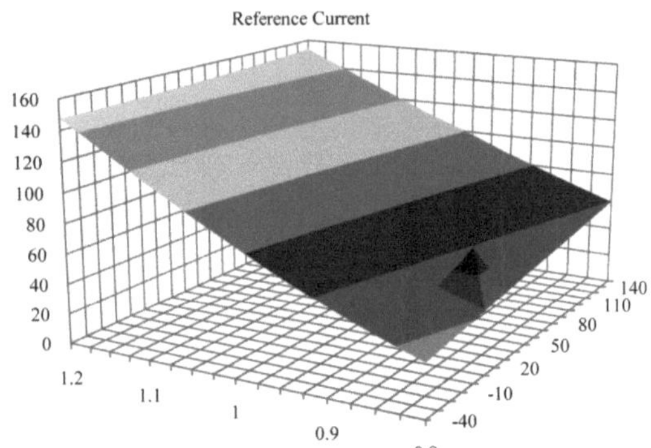

Fig. 8.7 *Iref* versus supply voltage and temperature

8.2 Post-Manufacture Variation Adaptation

Process variations may cause signal integrity problems in a current sensing interconnect. This negatively impacts the manufacturing yield. Techniques are needed to alleviate the effects of such variations. The traditional assumption of worst-case variation and guard-banding technique which uses large current margins has high power consumption costs. The other approaches can be classified into two: circuit optimization techniques such as V_t modulation, and post-manufacture circuit tuning techniques. In [138] post-manufacture variation adaptation technique to keep the delay and leakage power of the circuit within an acceptable range has been proposed. This technique relies on a hardware framework that supports self-test and performs self-adaptation using optimization algorithms of design parameters.

The process variation tolerance technique proposed in this chapter also uses a post-manufacture self-adaptation mechanism. The receiver's input and reference current variations are the result of manufacturing fluctuations which are static. Wear-out and ageing also cause variation but they are time-dependent on the scale of months and years. Hence, in the proposed technique, interconnect's signal integrity test and calibration are performed at every power start-up of the system to tackle process, wear-out and aging related variations. If an error is detected then the receiver, driver or both are reconfigured according to the developed algorithm and methodology. This makes the link adaptive to the effect of variations and thus enabling continuous and reliable operation of the interconnect. It also results in lower power consumption when compared to the worst-case approach.

8.3 Calibration for Process Variation Tolerance

The post-manufacture calibration technique has two advantages. The first and the most important one is ensuring tolerance to process variation and reliable communication by making the link adaptive to the effects of variation. The second one is reducing power consumption. Rather than assuming worst-cases and allocating large current margin which causes unnecessary power consumption, the margin is adjusted at every power start-up during the calibration phase by detecting the existing amount of variation. Based on the detection, receiver and driver reconfiguration will be performed. This is an efficient technique since it saves power by optimizing the margin and at the same time guarantees reliability. An error detection scheme as well as reconfiguration algorithms and methodology are developed. Furthermore, reconfiguration control and communication circuits are designed and simulated for a multilevel current sensing interconnect.

8.3.1 Algorithm and Methodology

The interconnect's signal integrity test is initiated by the receiver. When error is detected, receiver reconfiguration will be carried out first; if it is not enough to handle the variation then driver reconfiguration will be followed. Upon successful completion of calibration processes, the interconnect will be ready for data transmission phase. However, if both reconfigurations are failed the link will be declared as 'faulty/do not use'. The flows of the interconnect calibration process are shown in Fig. 8.8.

The calibration process is formulated by considering a three-level current sensing interconnect (0, I_1, and I_2) the same as in *PMCmFCD* and *DualdiffFCD* interconnects. The reason for choosing three-level is that *PMCmFCD* and *DualdiffFCD* interconnects have superior performance and better power efficiency than the *LEDRCm* which uses binary current sensing signaling (see Chap. 7). In fact, the formulated calibration process is scalable to any current sensing interconnects including binary current sensing signaling (0 and I). In a three-level current sensing interconnect Equations 8.6 to 8.8 should be satisfied in order to ensure its signal integrity. Besides data wires, four additional wires are needed to carry out the calibration (Fig. 8.9). *Calib_Ack*, *Calib_Req*, *C1* and *C2* wires are needed for handshaking and communicating the results of reconfigurations between sending and receiving modules during the calibration phase.

$$I_{refl} < I_1 < I_{ref2} \tag{8.6}$$

$$I_2 > I_{ref2} \tag{8.7}$$

$$I_{ref2} > I_{ref1} \tag{8.8}$$

Reconfigurable driver and receiver current sources of a three-level current sensing interconnect are shown in Fig. 8.10. When the receiver is ready to accept data it closes the switch of *Imin2* and *Imin1*. These switches stay closed all the time because to achieve the minimum required performance I_1 and I_2 wire currents must be greater than *Imin1* and *Imin2* currents, respectively. From Equation 8.6 and 8.8, it can be deduced that I_2 is always greater than I_1 which in turn means that

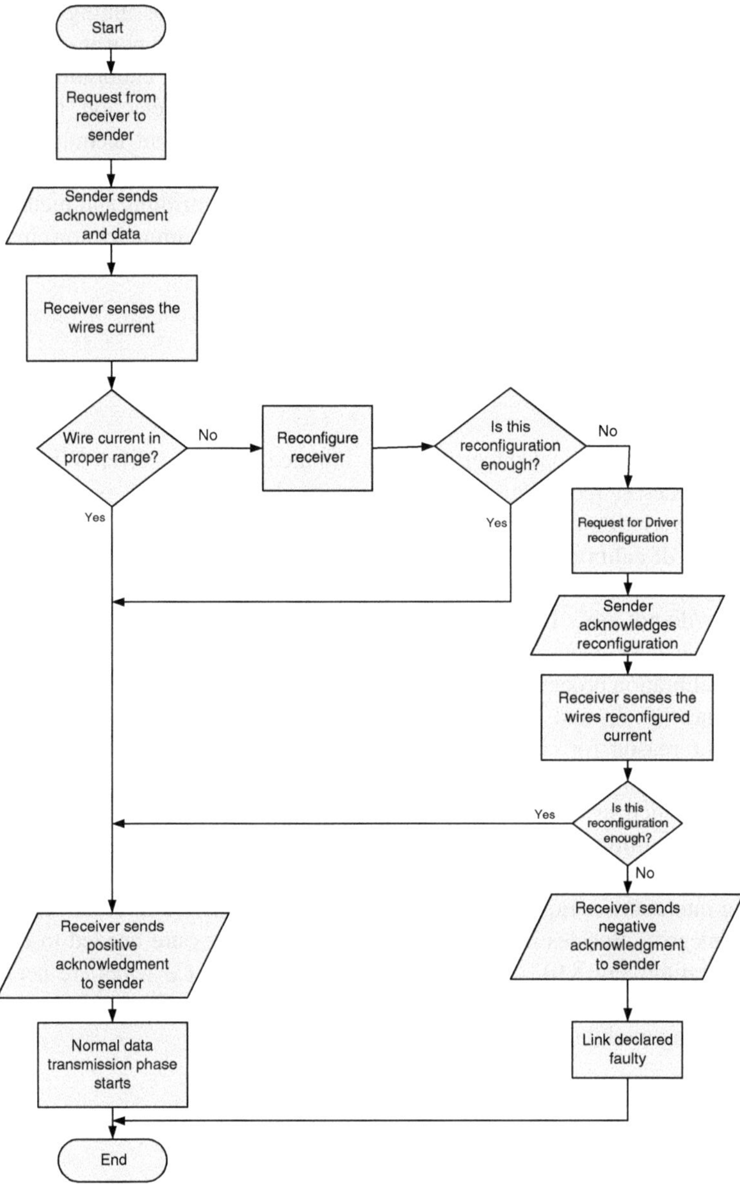

Fig. 8.8 Interconnect calibration flow chart

Imin2 is greater than *Imin1*. Initially the switches of *Ivar1* and *Ivar2* will also be closed. These two switches are needed to tolerate I_1 and I_2 wire current variation respectively. Depending on the variations, these two switches might be opened as a result of receiver reconfiguration. Additional current source *Ivar21* is required in order to avoid an error wherein a considerable increase (variation) in I_1 could

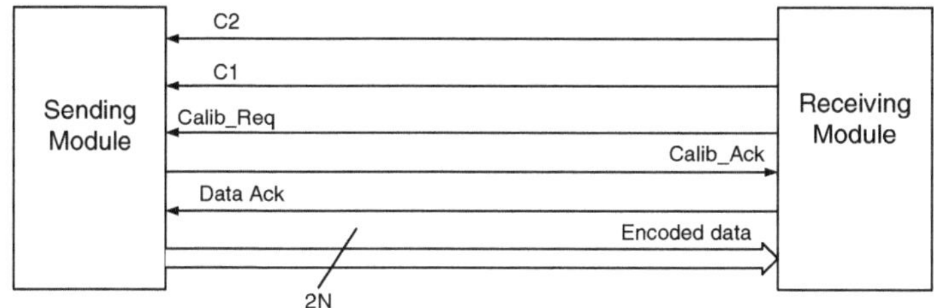

Fig. 8.9 Interconnect with calibration wires

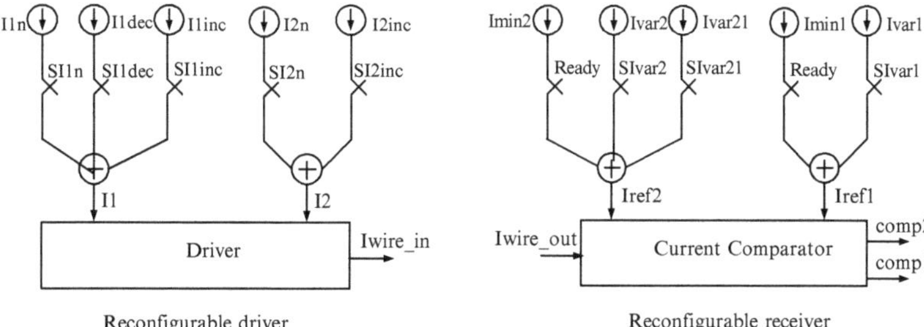

Fig. 8.10 Reconfigurable driver and receiver

mislead the receiver to interpret it as I_2. In this case I_{ref2} will also be increased. If I_1 increases it is more likely that I_2 also increases, because the drivers of I_1 and I_2 are placed closer to each other making them highly spatially correlated. If I_2 is not increased like in I_1 then the driver reconfiguration can be used to increase it, if necessary. Initially the switch of *Ivar21* is open and will be closed when needed.

The receiver's output signals are *comp1* and *comp2*. In a reliable interconnect, when I_1 is being transmitted, *comp1* should be *high* and *comp2* should be *low*. When I_2 is transmitted both *comp1* and *comp2* should be *high*. These two cases will be checked for all wires during the calibration process. In the driver, the switches *SI1n* and *SI2n* are always closed as they are the minimum currents that are required to achieve the desired communication performance. Initially, switch *SI1dec* is also closed and will be opened if the receiver requests for decreasing I_1. Switches *SI1inc* and *SI2inc* are open and will be closed when necessary.

The calibration process will be initiated by the receiver when it sends a request through *Calib_Req* wire to the sender. The sender sends an acknowledgment through *Calib_Ack* wire and current I_1 through all data wires immediately after it gets the first request transition (low-to-high). Outputs *comp1* and *comp2* of all wires will be checked after the acknowledgment signal arrives at the receiver. The algorithm for I_1 calibration is presented as pseudocode in Algorithm 8.1. There are three possible scenarios depending on the amount of variation. In *Case 1*, I_1 is in proper range. In *Case 2*, I_1 becomes less than expected and receiver cannot detect it. Finally in *Case 3*, I_1 becomes much larger than expected and receiver might detect it as I_2.

Algorithm 8.1 Calibration of I_1

```
Receiver: Req = 0 → 1; // request for start of calibration;
Sender:   Ack = 0 → 1 and transmit I1 through data wires;
Receiver: gets acknowledgment;
Case 1:
Receiver: if (comp1 = 1 and comp2 = 0) then
          reconfiguration is not required;
          Req = 1 → 0; // I1 is reliable and request for I2
Case 2:
Receiver: if (comp1 = 0 and comp2 = 0) then
           opens Ivar1 switch;
           check comp1 and comp2;
           if (comp1 = 1 and comp2 = 0) then
              receiver reconfiguration is successful;
              Req = 1 → 0; // I1 is reliable and request for I2
           else          // Receiver reconfiguration is not sufficient
              sends a pulse on C1 wire; // request for I1 increase
Sender:   Ack = 1 → 0;          // sends acknowledgment
Receiver: gets acknowledgment;
        check comp1 and comp2;
        if (comp1 = 1 and comp2 = 0) then
          driver reconfiguration successful;
          Req = 1→0; // I1 is reliable and request for I2
        else
           sends a second pulse on C1; // I1 calibration not successful
Case 3:
Receiver: if (comp1= 1 and comp2 = 1) then
            close switch of Ivar21;
            check comp1 and comp2;
            if (comp1 = 1 and comp2 = 0) then
              receiver reconfiguration successful;
              Req = 1 → 0;   // I1 is reliable and request for I2
            else
              sends pulse on C1 and C2   // request for I1 decrease
Sender:   Ack = 1 → 0;   // sends acknowledgment
Receiver: gets acknowledgment;
        check comp1 and comp2;
        if (comp1 = 1 and comp2 = 0) then
           driver reconfiguration is successful;
           Req = 1 → 0;    //I1 is reliable and request for I2
        else
         sends a second pulse on C1    // I1 calibration not successful
```

Upon getting a high-to-low request signal transition, the sender drives I_2 through data wires and acknowledgment through *Calib_Ack* wire. The algorithm for I_2 calibration is presented as pseudocode in Algorithm 8.2. There are also three possible variation cases for I_2. In *Case 1*, I_2 is in proper range. In *Case 2*, I_2 is less than expected and this causes error because the receiver can detect it as I_1. In the last case, I_2 may even be less than I_1 and the receiver cannot detect any transmission in this case.

Algorithm 8.2 Calibration of I_2

```
Sender:   Ack = 0 → 1 or 1 →0;     // depending on previous state
          transmits I2 ;
Receiver: gets acknowledgment;
Case 1:
Receiver: if (comp1 = 1 and comp2 = 1)  then
          reconfiguration is not needed;
          Req = 0 → 1;    // I2 is in proper range and calibration
                             completed successfully and link is reliable
Case 2:
Receiver: if (comp1 = 1 and comp2 = 0)  then
          opens Ivar2 switch;    // receiver reconfiguration
          check comp1 and comp2;
          if (comp1 = 1 and comp2 = 1)  then
            Req = 0 → 1;   // calibration completed successfully
          else
           sends a pulse on C2 wire;    // request for I2 increase
Sender:   Ack = transition; // sends acknowledgment
Receiver: gets acknowledgment;
        check comp1 and comp2;
        if (comp1 = 1 and comp2 = 1)  then
          Req = 0 → 1;    // calibration completed successfully
        else
          sends a second pulse on C2;    // calibration not successful
Case 3:
Receiver: if (comp1 = 0 and comp2 = 0)  then
          sends a pulse on C2;    // request for I2 increase
Sender:   Ack = transition;    // sends acknowledgment
Receiver: gets acknowledgment;
        check comp1 and comp2;
        if (comp1 = 1 and comp2 = 1)  then
          Req = 0 → 1;    // calibration completed successfully
        elseif (comp1 = 1 and comp2 = 0)  then
         opens Ivar2 switch;
         check comp1 and comp2;
         if (comp1 = 1 and comp2 = 1)  then
          Req = 0 → 1;    // calibration completed successfully
         else
          sends a second pulse on C2;    // calibration not successful
```

The success of calibration and the links reliability is confirmed when there is a second low-to-high request transition through *Calib_Req* wire. A calibration failure is indicated by sending a second pulse either on *C1* or *C2* wire. The calibration process can be classified into best, average, worst and failure cases depending on the number of steps required and the final result (successful or failure). In the best case either there is no need for reconfiguration at all, or only receiver reconfiguration is enough (Fig. 8.11). In the average case only one driver reconfiguration either on I_1 or I_2 besides receiver reconfiguration is required. There are three possible ways in which the average case can be manipulated: by increasing I_1, decreasing I_1 and increasing I_2 (Fig. 8.12).

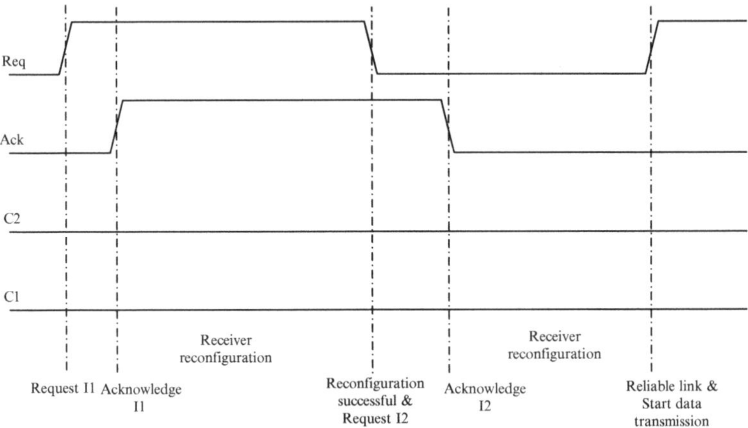

Fig. 8.11 Best case: driver reconfiguration is not required

In the worst case, two driver reconfigurations are needed in addition to the receiver reconfiguration (Fig. 8.13). There are two possible ways in which the worst case can be handled: by increasing both I_1 and I_2 and decreasing I_1 and increasing I_2. If both receiver and driver reconfigurations failed to make the link adaptive to the variation, then the link is in failure state (Fig. 8.14). At this stage the error will be reported to a higher level error controlling system. This leads to a more power efficient error detection and correction scheme, because higher level error controlling mechanisms will come into play only when necessary. There are three failure scenarios: does not compensate for the variation by decreasing/increasing I_1 or increasing I_2 besides receiver reconfigurations. Upon completion of the calibration process successfully, data transmission phase will start.

8.3.2 Reconfiguration Control and Communication Circuits

In general the circuits can be classified into two parts: driver and receiver side circuits. The receiver side circuit detects the receiver's outputs and performs the needed reconfiguration at the receiver by increasing or decreasing reference currents. It also sends requests for I_1 and I_2 transmissions, and for driver reconfiguration when required. In addition, it communicates the calibration results with the transmitter. The driver sends either the nominal or reconfigured current through data wires and an acknowledgment signal through the *Calib_Ack* wire depending on the state of handshaking signals. The input and output signals of driver and receiver reconfiguration control circuits are shown in Fig. 8.15. The block level diagram is intended to provide a clear distinction between reconfiguration control and communication signals.

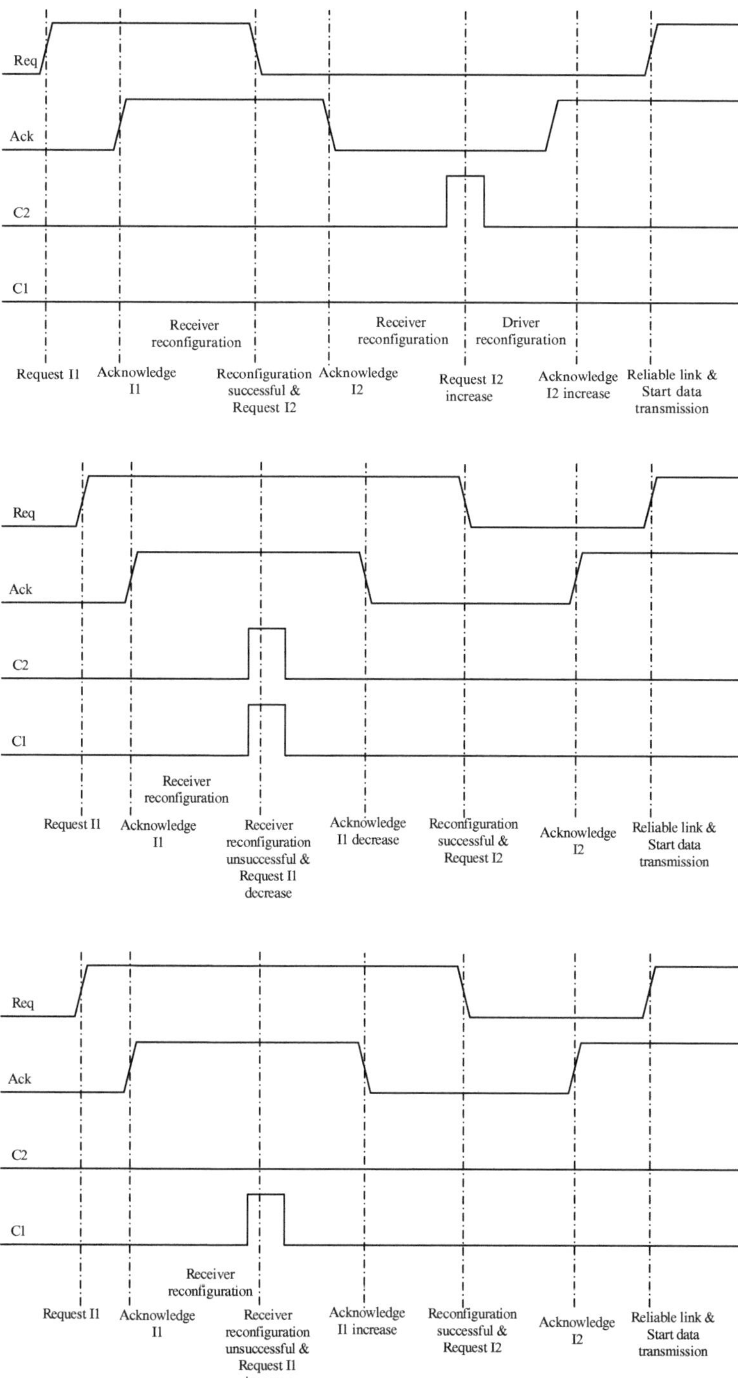

Fig. 8.12 Average case calibrations

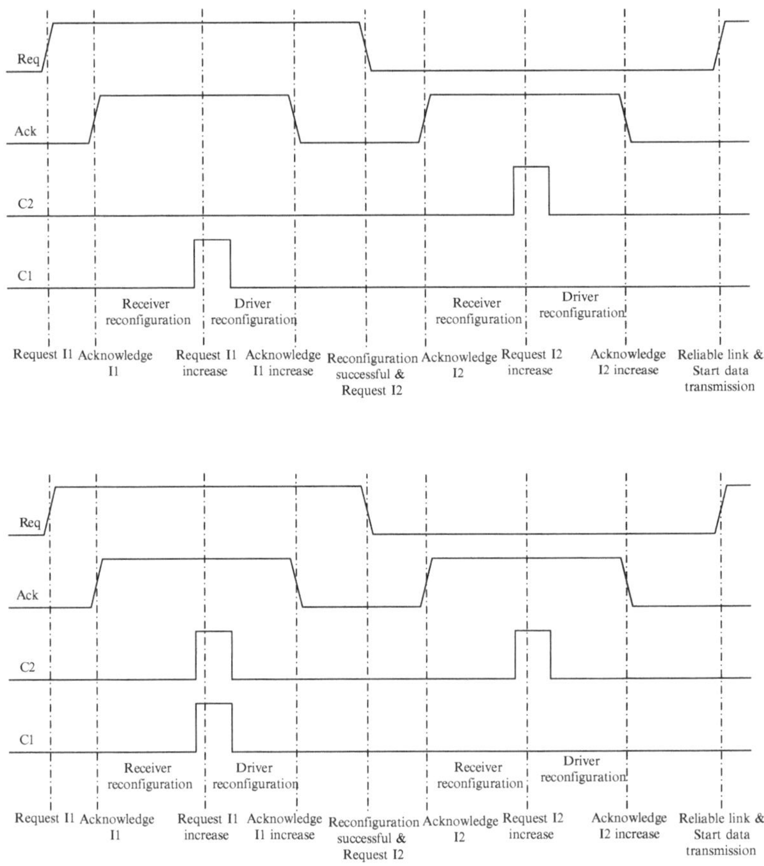

Fig. 8.13 Worst case calibrations

The receiver and driver reconfiguration control circuits are shown in Figs. 8.16 and 8.17, respectively. Immediately after power start-up, the receiver raises its *Ready* signal to *high* whenever it is ready to accept data, this makes *Reqin* signal to have a transition from low-to-high. Upon getting *Reqout* transition the sender transmits I_1 and *Ackin* (low-to-high transition) through data and *Calib_Ack* wires, respectively. The receiver checks its output signals, *comp1* and *comp2*, when it gets a transition in *Ackout*. Receiver reconfiguration will be performed by controlling the reference current's switches using *SIvar1*, *SIvar2* and *SIvar21* signals (Fig. 8.16). If the receiver reconfiguration is able to fix the signal integrity problem, then the receiver communicates it to the transmitter using *Reqin* signal transition. If not, it sends a request for driver reconfiguration using signals *C1pulse*, *C2pulse* or both depending on the required reconfiguration through their respective wires. Based on the receiver reconfiguration result, transmitter sends either I_2 or reconfigured I_1. The driver performs reconfiguration of I_1 using *SI1dec* or *SI1inc* signals (Fig. 8.17).

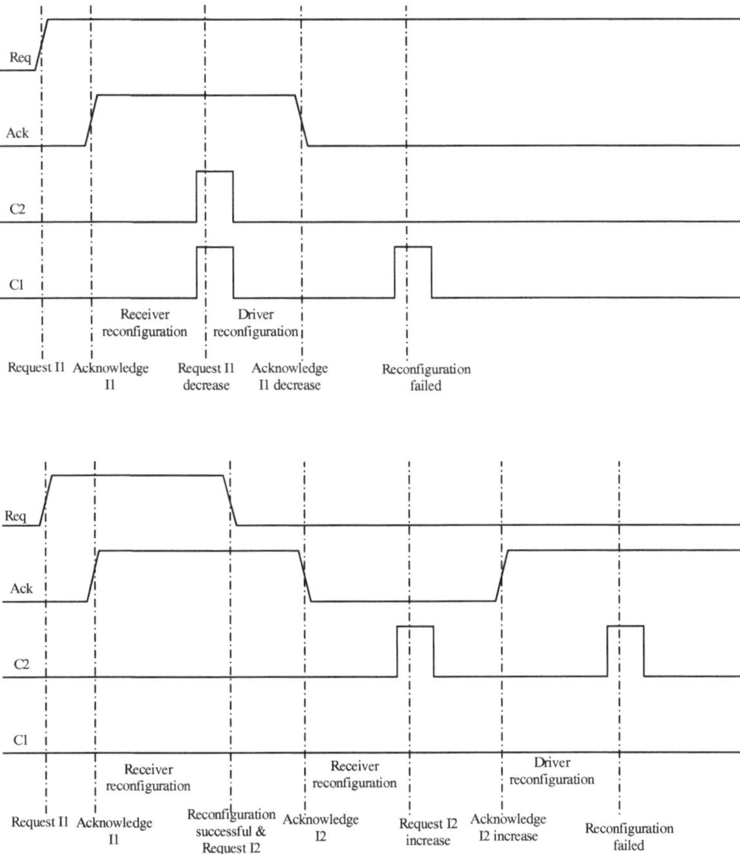

Fig. 8.14 Calibration failure cases

When the receiver requests for I_2 reconfiguration, it will do so using *SI2inc*. In case both sides of reconfigurations fail, a second pulse will be generated on *C1pulse* or *C2pulse* depending on the current level under the reconfiguration process. Upon getting a second pulse the transmitter raises *Calibration_Failed* signal to *high*, indicating the failure of the calibration to adapt with the variation. The success of calibration, in other words process variation tolerance of the interconnect is approved by the receiver when it generates a second low-to-high transition in *Reqin* signal (Fig. 8.16). The *Link_Reliable* signal will be *high* when the transmitter gets a second low-to-high transition on *Reqout* signal (Fig. 8.17).

The d(latch) and d(comp) in Fig. 8.16 are delay elements and their delay correspond to the delay of latch and current comparator, respectively. The *t2p* block in Fig. 8.16 and *p2t* block in Fig. 8.17 are transition-to-pulse and pulse-to-transition converters, respectively and their implementation is shown in Fig. 8.18.

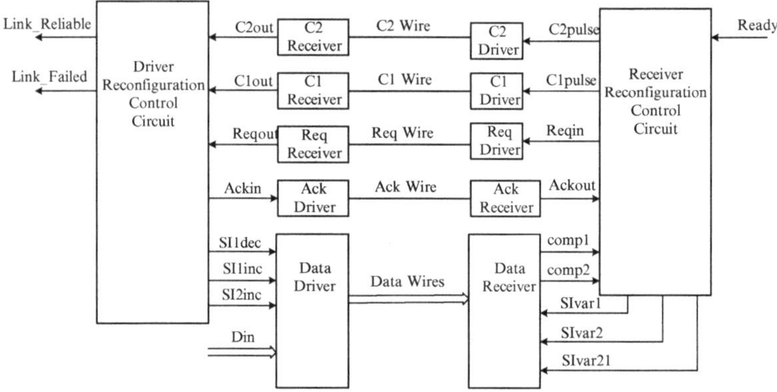

Fig. 8.15 Calibration control and communication signals

8.4 Runtime Management of Voltage and Temperature Variations

In a global on-chip communication link, transmitter and receiver are placed far apart from each other. At runtime, the supply VT of a transmitter can be different from the receiver depending on the spatial switching activities and hotspot localities. These in turn deviate the transmitter's output current, and consequently the receiver's input current from its nominal value. In a current sensing interconnect, where a receiver compares its input current with a reference current, this variation may affect reliability if it is out of the allocated margin. The usual trend is assuming worst-case variation and allocating large enough current margins between the receiver's input and reference currents which leads to additional power consumption.

An alternative power efficient technique, is monitoring the variation at runtime and adjusting the interconnect circuits when a signal integrity problem is detected. This enables power efficient, runtime error detecting and correcting scheme. To do so, circuit level variation sensing mechanism along with its implementation is devised and presented in this section. When an error is detected due to runtime variation, reconfiguration of the interconnect circuits and retransmission of the data will be carried out.

8.4.1 Sensing Effects of Voltage and Temperature Variation

Sensing is an important task of any adaptive system that compensates for variation. A sensor monitors the runtime operating conditions of a system. Here runtime variation of $I_{rec,in}$ along with receiver's reference current will be monitored. If the variation causes error, the error will be reported to both the transmitter and receiver, besides reconfiguring the receiver to adapt with the variation. The receiver output

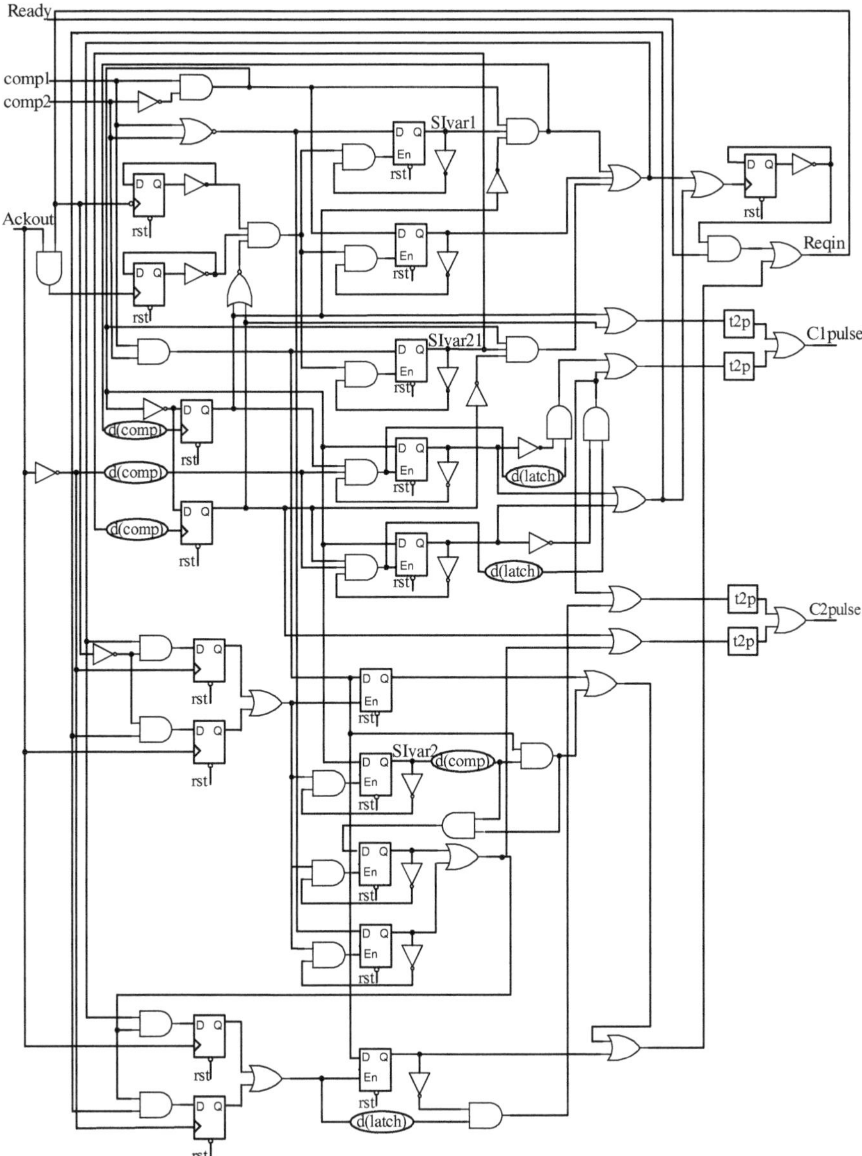

Fig. 8.16 Receiver reconfiguration control circuit

will be erroneous if $I_{rec,in}$ becomes lower than expected or receiver's reference current increases more than needed. There are three causes for the error: large supply voltage drop at the transmitter side, significant increase in temperature at the receiver or both. So, effects of supply voltage fluctuation at the transmitter and temperature variation at the receiver are considered in the design of the sensing circuit.

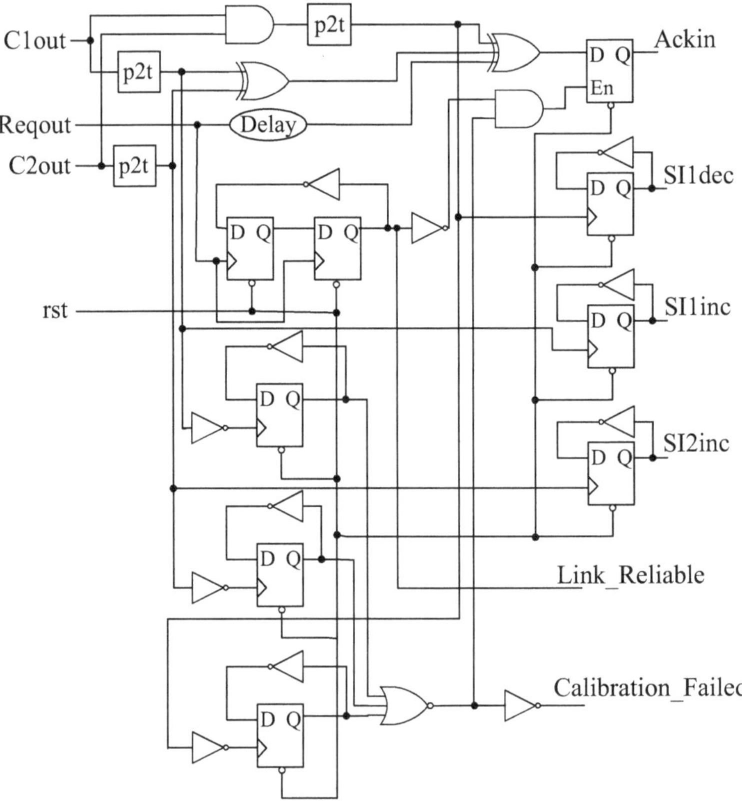

Fig. 8.17 Driver side calibration control

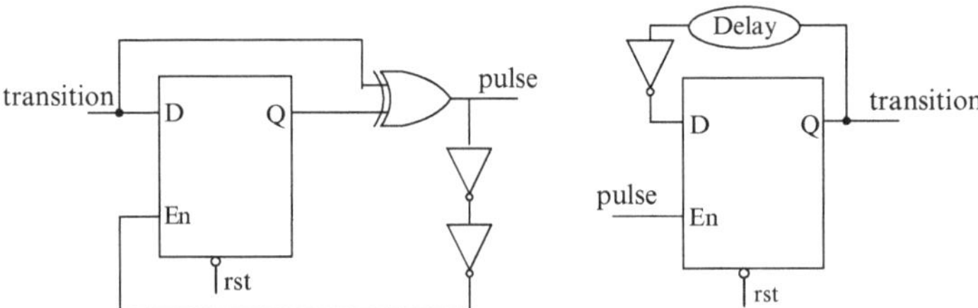

Fig. 8.18 Transition-to-pulse and pulse-to-transition converters

Runtime monitoring of VT variations is carried out using two additional wires which run adjacent to data transmission wires (Fig. 8.19). With every new data transmission, the same amount of current as in data wires will be transmitted through the sensing wires by changing the direction of current. The sensor circuit at the receiver compares the sensing wire's current with the receiver's reference current. If the sensing wire's current is greater than the reference, the sensor output stays low,

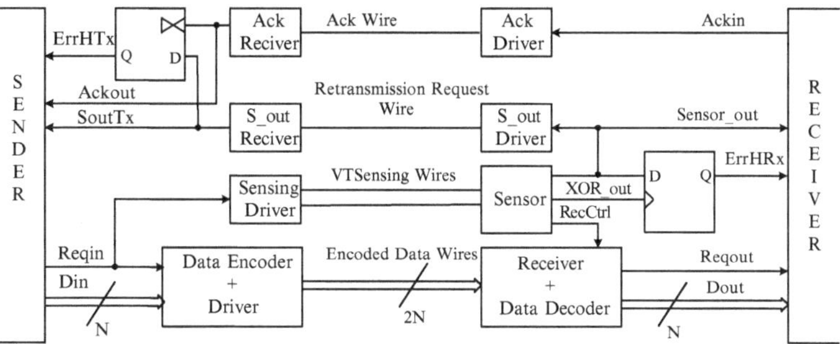

Fig. 8.19 Interconnect with VT runtime variations detector

indicating there is no variation which causes error. If the sensing wire's current is less than the reference, then sensor output goes high, thus detecting error.

8.4.2 Sensor Circuit Implementation

The sensor circuit is based on current subtraction. It subtracts the receiver's reference current from the sensing wire's current. The sensor circuit is shown in Fig. 8.20. The current direction changes in the sensing wires with every new data transmission in the channel. For example, if I_{vts1} flows towards the receiver, then I_{vts2} flows towards the transmitter and vice versa in the next transmission. Due to this, there is current either in *Mn1* or *Mn2* at any time. Current I_{vts1} and I_{vts2} are mirrored to *Mn5* and *Mn6*, respectively. The reference current is mirrored to *Mp4*. The current comparator, which is based on current subtraction, compares I_{ref} with either I_{vts1} or I_{vts2} depending on the one which flows towards the receiver. The comparator output is buffered to make the *Cout* signal full swing.

The purpose of the current direction sensor is to know the arrival of new data and check for its reliability when it is valid and stable. The current direction sensor circuit is shown in Fig. 8.20 and was used as part of a receiver in *Dualdiff* interconnect. Consider the top current direction sensor in Fig. 8.20. Transistor *Mp1* provides negative feedback to transistor *Mn3*. It switches the gate of *Mn3* on and off as required and helps in modulating the input impedance. Transistor *Mp2* provides a constant current bias and thus regulates the transconductance of *Mn3*. The source terminal of transistor *Mn3* is connected to the I_{vts1} wire. When current flows towards the driver, *Mn3* switches to on state and pulls the output of the current sensor to *low*. When current is sourced by the driver, the source voltage of *Mn3* rises thus switching it off. In this case, current flows through the load transistor *Mn4* to the output, making the output voltage of the current direction sensor *high*. The output of the two direction sensors are used as inputs to the XOR gate. The additional delay due to XOR gate and buffers, ensures the stability of *Cout* before *XOR_out*

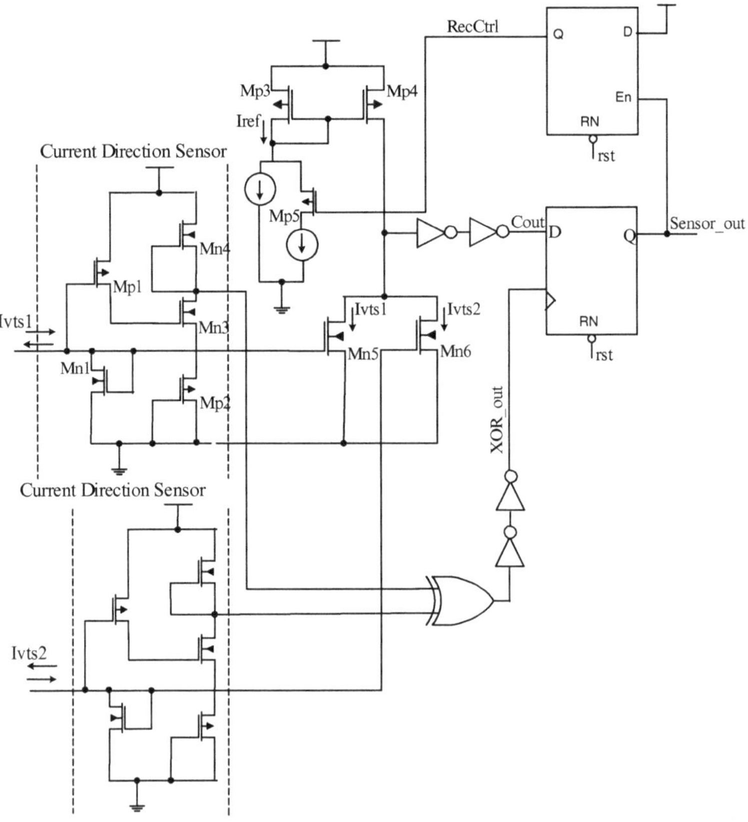

Fig. 8.20 VT variation sensor and reconfiguration circuit

becomes *high*. If either I_{vts1} or I_{vts2} is less than I_{ref}, then *Cout* becomes *high*. This output will be latched to *Sensor_out* when *XOR_out* makes a transition to *high*, thus detecting the error due to runtime VT variations.

8.4.3 Reconfiguration and Retransmission

The receiving block always checks the output of the sensor circuit *Sensor_out* before it uses or forwards the received data. When *Sensor_out* becomes *high* it suspends the use of data until *Sensor_out* returns to *low*. Reconfiguration of the receiver, more specifically decreasing the reference current *Iref*, will be carried out when the sensor detects error. The latch will be enabled when *Sensor_out* is *high* which in turn makes *RecCtrl high* and then switches *Mp5* to a non-conducting state. In order to check the success of reconfiguration in withstanding the variation effect, retransmission request will be sent to the transmitter through the retransmission request wire (Fig. 8.19). *Sensor_out* signal is used as a retransmission request signal. When the transmitter gets a retransmission request signal, it sends the data again and changes the current directions in the sensing wires. *XOR_out* signal makes a transition to *high* due to current direction changes in sensing wires, allowing *Cout* to be latched to *Sensor_out*. If the reconfiguration is successful, *Sensor_out* becomes *low* and the receiver and transmitter resume their normal data transmission phase. Otherwise, the error will be reported to the transmitter, receiver and the higher level error controlling system. When the reconfiguration fails, *ErrHRx* signal becomes *high*, and then the receiver sends negative acknowledgment through *Ackwire*, which in turn makes *ErrHTx high*, thereby informing the failure of reconfiguration to the transmitter. The transmitter translates *Ackout* signal transition as negative acknowledgment when *SoutTx* is *high* (Fig. 8.19). If there is a transition in *Ackout* signal and *SoutTx* is *low* then the transmitter sends the next data. Whereas it retransmits the same data if there is no transition in *Ackout* and *SoutTx* is *high*. When the reconfiguration has failed to tolerate the effect of variations, the link will be labeled as temporarily failed until the higher level error controlling system fixes the variation, for instance by decreasing the switching activities to make the voltage drop in an acceptable margin.

8.5 Simulation Results and Analysis

Simulations of *PMCmFCD* and *DualdiffFCD* interconnects consisting of the calibration and VT runtime variation management circuits were designed and performed in Cadence Analog Spectre using 65 nm CMOS technology from STMicroelectronics and 1V supply voltage. The interconnect length was set to 2 mm, the same as in inter router link length of Intel 80-Tile TeraFLOPS processor [69]. The wire properties were set according to ITRS 65 nm technology node for global wiring. The RLC values of the wires were extracted using field solvers for microstrip configuration. The resistance and inductance values were extracted using FastHenry [45], while the capacitance values were extracted using Linpar [46]. In calibration wires (*Calib_Req*, *Calib_Ack*, *C1* and *C2*) voltage-mode signaling with repeater insertion was used.

The time taken and average power consumed during the calibration process is listed in Table 8.1. The calibration delay for best-case, where driver reconfiguration was not required, was 2.66 ns. This is the minimum delay incurred due to calibration at every power start-up of the system. The best-case requires five communications between the sender and the receiver. They are separate request and acknowledgment for both I_1 and I_2 transmission besides communicating the robustness of the link. The average-case delay, which requires one driver reconfiguration in addition to receiver reconfiguration, was 4.19 ns. The delay of the worst-case calibration was 5.72 ns, which requires both driver and receiver side reconfiguration for I_1 and I_2. The average power consumed during the calibration process is low. However, its peak power is high, (Table 8.1) but it occurs only for a very short period.

Table 8.1 Calibration delay and power consumption

Classification	Delay [ns]	Average power [μW]	Peak power [μW]
Best-case	2.66	164	1188
Average-case	4.19	179	1392
Worst-case	5.72	243	1345

Table 8.2 Calibration area overhead

Bit width [bits]	Active area overhead [%]	Wiring area overhead [%]
2	50.91	38.88
4	38.96	26.92
8	26.23	16.66
16	15.76	9.72
32	8.80	5.07
64	4.67	2.63

In order to examine the power saving benefits of the presented calibration technique, three 64-bits wide *PMCmFCD* interconnects consisting of the calibration circuits were designed. The first interconnect was designed by guard-banding for worst-case variation. Let us quantify it as $\pm 3\sigma$ variation which accounts for about 99.7% of the overall variation range. The second interconnect is designed by allocating current margins for $\pm 2\sigma$ variations, which covers about 95.4% of the variation. The third one is designed with current margins for $\pm 1.5\sigma$ variations. The second and third interconnects have saved 7.88% and 14.21% power, respectively over the interconnect with worst-case margin allocation. This proves that allocating smaller margins and relying on the proposed calibration technique will lead to a better power efficiency than the conventional worst-case design.

The required additional area due to the calibration circuits was determined from the Cadence schematic. It requires $25\mu m^2$ active and $2940\mu m^2$ wiring areas. The calibration area overhead for *PMCmFCD* interconnect has been calculated and it decreases for larger bit width transmissions (Table 8.2). For example, the active area overhead for 4-bits and 64-bits *PMCmFCD* interconnects are 38.96% and 4.67%, respectively.

Simulation waveforms for an average-case calibration process is shown in Fig. 8.21. As it can be seen from the simulation waveforms, *Reqin* signal goes *high* when it gets *Ready* signal from the receiver and *Ackin* goes *high* after it gets a *high* transition on the *Reqout* signal. The receiver's reconfiguration control checks *comp1* and *comp2* and detects error, because both *comp1* and *comp2* were *low*. Then it reconfigures the receiver by turning on the switch of *Ivar1*, making *SIvar1* signal *high* by decreasing I_{ref1}. But this is not enough to control the effect of variation. Then the receiver requests for an increase in I_1 by sending *C1pulse*. Upon getting a pulse on *C1out*, the transmitter reconfigures the driver, increases I_1 by closing switch *SI1inc*. It also transmits the reconfigured I_1 along with acknowledgment. The receiver checks *comp1* and *comp2* signals when *Ackout* makes transitions to *low*. Then both the signals are in proper range, which confirms the reliability

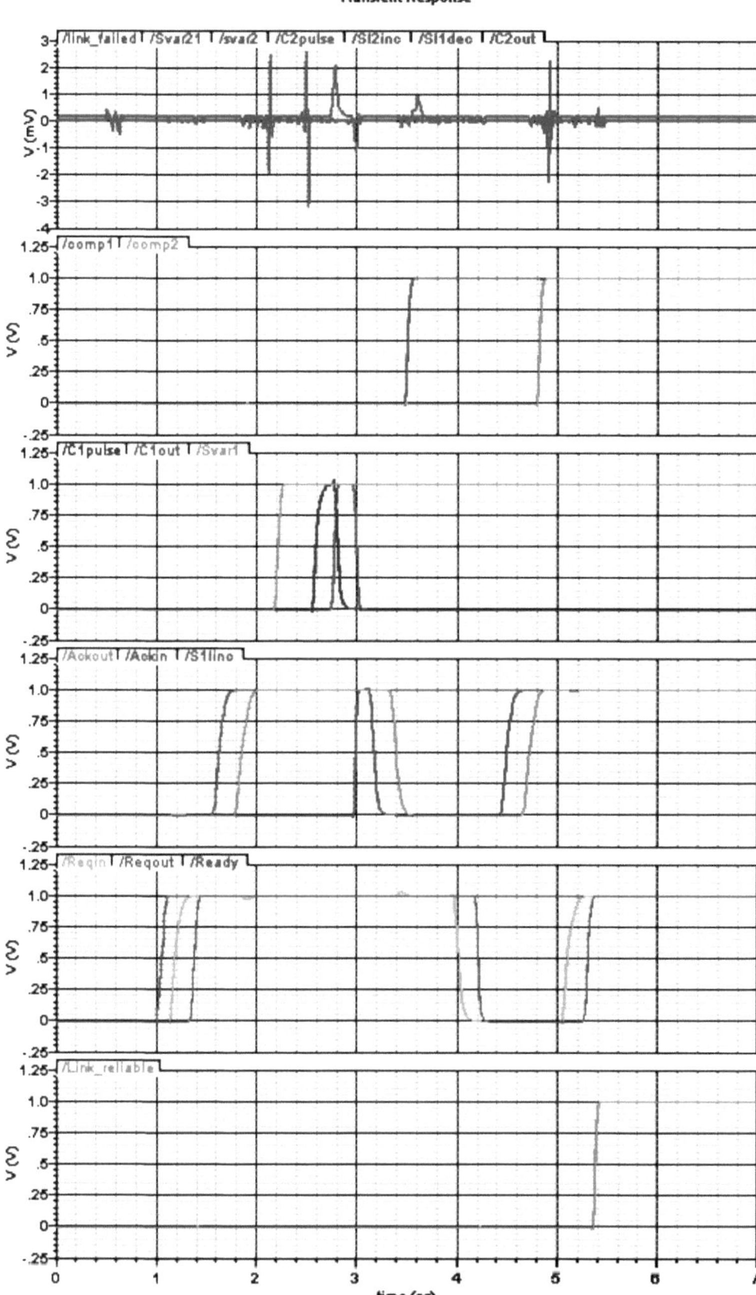

Fig. 8.21 Simulation waveforms of average-case calibration

Table 8.3 Power overhead of VT variation management

Bit width [bits]	PMCmFCD [%]	DualdiffFCD [%]
2	26.8	26.3
4	21.6	17.2
8	10.9	10.2
16	6.2	5.6
32	3.2	2.9
64	1.6	1.5

Table 8.4 VT variation management area overhead

Bit width [bits]	Active area overhead [%]	Wiring area overhead [%]
2	19.47	31.24
4	12.95	20.83
8	7.65	12.49
16	4.18	7.14
32	2.20	3.67
64	1.13	1.89

of I_1 transmission. The receiver requests for I_2 transmission by sending a high-to-low transition in *Calib_Req* wire. When *Reqout* makes a transition from high-to-low, the transmitter sends signals I_2 and acknowledgment (transition to *high*). The receiver checks signals *comp1* and *comp2* when it gets a transition in *Ackout*, both are *high*, indicating I_2 is in proper range. The receiver then declares the interconnects reliability by sending a low-to-high transition in *Calib_Req* wire. The transmitter declares the link's reliability upon getting a second low-to-high transition on *Reqout* signal by raising the *Link_reliable* signal to *high*.

After the calibration phase is completed, normal data transmission along with runtime VT variation sensing is performed. The delay of the VT variation sensor circuit was measured and it is 184 ps. The power consumed due to monitoring the variation was 489 μW. This is the only power overhead if there is no retransmission due to error. As can be seen from Table 8.3, this power overhead decreases for larger bit width transmissions. For example, in *DualdiffFCD* interconnect 4 and 64-bit transmissions the power overheads are 17.2% and 1.5%, respectively. The power overhead is reasonable and affordable for 16-bits and wider transmissions.

It is possible to have power saving rather than power overhead using the proposed VT management circuits. For example, instead of the usual worst-case guardbanding for 10% V_{DD} variation (which is equivalent to 3σ according to the ITRS roadmap [63]), allocating margin for 6.67% V_{DD} variation covering 95% of the variations range can be sufficient. The rest will be relied on the proposed runtime monitoring and reconfiguration. This approach results in 2% power savings in a 64-bit *PMCmFCD* interconnect instead of consuming additional power.

The required active and wiring areas for this technique (including retransmission and reporting error to higher level error controlling system) are $5.83\mu m^2$ and $2100\mu m^2$, respectively. The portion of area taken by the VT management circuits have been determined for *PMCmFCD* interconnect and listed in Table 8.4. In 64-bit

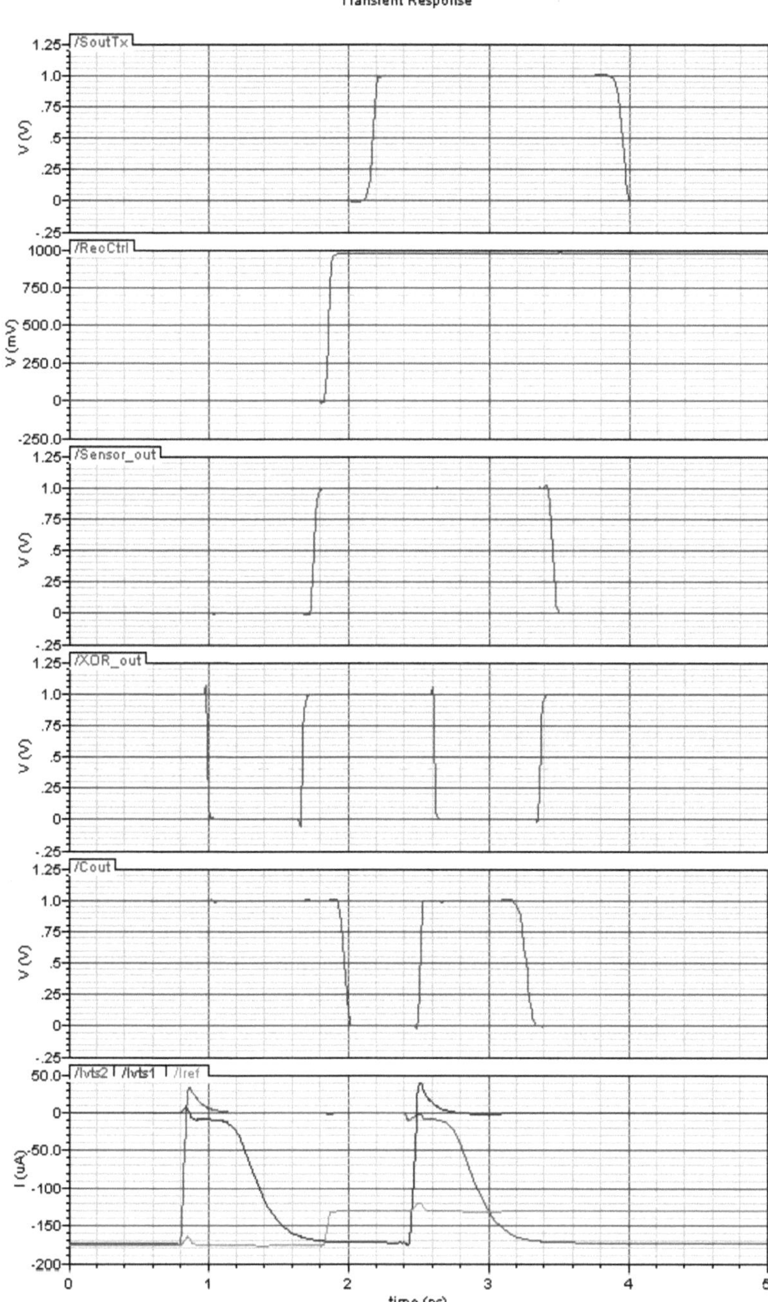

Fig. 8.22 Simulation waveforms of VT variation tolerance

PMCmFCD interconnect, it requires only 1.13% of the active area and 3.81% of the wiring area. The area overhead becomes smaller for larger bit width transmissions, thus showing its appropriateness for real life applications.

Simulation waveforms are shown in Fig. 8.22 for a nominal supply voltage of 1V, and assuming that there is a supply voltage drop at the transmitter which causes receiver's input current variation that leads to an error. The sensor then detects an error and it is flagged by making *Sensor_out high*. This in turn leads to reconfiguration of the receiver by making *RecCtrl* signal *high* and at the same time a retransmission request is sent to the transmitter, by making *SoutTx* signal *high*. Sensing after reconfiguration and arrival of the retransmitted data asserts the reliability of the link by making *Sensor_out* signal *low*.

8.6 Chapter Summary

In this chapter, circuit techniques for PVT variations tolerance for current sensing on-chip interconnects are presented. The technique for process variation tolerance is based on detecting signal integrity of the interconnect and performing calibration at every power start-up of the system. If an error is detected, the receiver's reference and/or input currents will be adjusted through receiver and driver reconfigurations. This makes the interconnect adaptive to the effects of process, wear-out and aging related variations, thereby enabling its continuous and reliable operation.

In a current sensing interconnect, using traditional worst-case guardbanding to tolerate environmental variations is costly and may not even be sufficient as the amount of runtime variation in high performance ICs are increasing. This makes runtime VT variation management technique an alternative and a better approach. The presented technique for runtime VT variations tolerance is based on monitoring the effect of their runtime variations and then reconfiguring the receiver when an error is detected. After the reconfiguration, request for data retransmission will be sent. The power and area overhead of this technique is low especially for larger bit width transmissions. It has been even proved that power can be saved compared to the worst-case approach.

References

1. S. Borkar, *Design challenges of technology scaling*. Micro, IEEE, 19(4):23-29, 1999.
2. G. E. Moore, *Cramming More Components on Integrated Circuits*. Electronics (38), 8:114-117, 1965.
3. Karl Goser, Peter Glsektter and Jan Dienstuhl, *Nanoelectronics and Nanosystems: From Transistors to Molecular and Quantum Devices*. Springer-Verlag Berlin, 2004.
4. R. Kumar and G. Hinton, *A Family of 45nm IA Processors*. 2009 IEEE International Solid-State Circuits Conference, Digest of Technical Papers, Vol. 1, pp. 58-59, Feb. 2009.
5. S. Rusu, S. Tam, H. Muljono, D. Ayers, J. Chang, R. Varada, M. Ratta, and S. Vora, *A 45nm 8-Core Enterprise Xeon® Processor*. 2007 IEEE Asian Solid-State Circuits Conference, pp. 9-12, Dec. 2009.
6. T. Sakurai, *Perspectives on power-aware electronics*. 2003 IEEE International Solid-State Circuits Conference, Digest of Technical Papers, Vol. 1, pp. 26-29.
7. N. Magen, A. Kolodny, U. Weiser, and N. Shamir, *Interconnect power dissipation in a microprocessor*. in IEEE/ACM International Workshop on System Level Interconnect Prediction, pp. 7-13, Feb. 2004.
8. S. Vangal, A. Singh, J. Howard, S. Dighe, N. Borkar, and A. Alvand-pour, *A 5.1GHz 0.34mm^2 router for network-on-chip applications*. in Symposium VLSI Circuits Dig. Tech. Papers, pp. 42-43, Jun. 2007.
9. B. P. Wong, F. Zach, V. Moroz, A. Mittal, G. W. Starr, and A. Kahng, *Nano-CMOS Design for Manufacturability*. A John Wiley & Sons Inc Publication, 2009.
10. M. Orshansky, S. Nassif, D. Boning, *Design for Manufacturability and Statistical Design: A Constructive Approach*. Springer, 2007.
11. Dennis Sylvestera, Kanak Agarwalb and Saumil Shaha, *Variability in nanometer CMOS: Impact, analysis, and minimization*. Integration, the VLSI Journal, Vol. 41, No. 3, pp. 319-339, May 2008.
12. S. R. Nassif, *Model to Hardware Matching; For nano-meter Scale Technologies*. 2006 International Conference on Simulation of Semiconductor Processes and Devices, pp. 5-8, Sept. 2006.
13. K. Bernstein, et al, *High-performance CMOS variability in the 65-nm regime and beyond*. IBM Journal of Research and Development archive, Vol. 50, No. 4/5, pp. 433-449, July 2006.
14. V. Agarwal, M. S. Hrishikesh, S. W. Keckler and D.Burger, *Clock rate versus IPC: the end of the road for conventional microarchitectures*. Proceedings of the 27th International Symposium on Computer Architecture, pp. 248-259, 2000.
15. P. Guerrier and A. Greiner, *A generic architecture for on-chip packet switched interconnections*. in Design, Automation and Test in Europe Conference and Exhibition, DATE'00, Paris, France, Mar. 2000, pp. 250-256.

E.E. Nigussie, *Variation Tolerant On-Chip Interconnects*, Analog Circuits and Signal Processing, DOI 10.1007/978-1-4614-0131-5,

16. A. Hemani et al, *Network on a chip: An architecture for billion transistor era.* in 18th NORCHIP Conference, Turku, Finland, Nov. 2000, pp. 166-173.
17. M. Sgroi et al, *Addressing the System-on-a-chip interconnect woes through communication-based design.* in 38th Design Automation Conference, DAC 2001, Las Vegas, NV, June 2001, pp. 667-672.
18. *TILE64^{TM} Processor. [Online].* Available http://www.tilera.com/products/TILE64.php
19. A. Upadhyay, S. R. Hasan, and M. Nekili, *A novel asynchronous wrapper using 1-of-4 data encoding and single-track handshaking.* The 2nd Annual IEEE Northeast Workshop on Circuits and Systems, pp. 205- 208, June, 2004.
20. M. Ferretti, and P. A. Beerel, *Single-Track Asynchronous Pipeline Templates Using 1-of-N Encoding.* Proceedings of the 2002 Design, Automation and Test in Europe Conference and Exhibition, pp. 1008-1015, August, 2002.
21. T. Hanyu, T. Takahashi and M. Kameyama, *Bidirectional data transfer based asynchronous VLSI system using multiple-valued current mode logic.* 33rd International Symposium on Multiple-Valued Logic, pp. 99-104, May 2003.
22. E. Nigussie, J. Plosila and J. Isoaho, *On Asynchronous Full-Duplex Dual-Rail Link with Multiple-Valued Current-Mode Signaling.* 23rd NORCHIP Conference, pp. 222-225, Nov. 2005.
23. E. Nigussie, J. Plosila and J. Isoaho, *Full-duplex link implementation using dual-rail encoding and multiple-valued current-mode logic.* 2006 IEEE International Symposium on Circuits and Systems ISCAS 2006, 4 pp, May. 2006.
24. K. Nabors, and J. White, *FastCap: a multipole accelerated 3-D capacitance extraction program.* 2006 IEEE Transactions on Computer-Aided Design of Integrated Circuits and Systems, Vol. 10, No. 11, pp. 1447-1459, Nov. 1991.
25. H. Ron, J. Gainsley and R. Drost, *Long wires and asynchronous control.* Proc. 10th IEEE International Symposium on ASYNC, pp. 240-249, Apr. 2004.
26. R. Dobkin, R. Ginosar, and C. P. Sotiriou, *High Rate Data Synchronization in GALS SoCs.* IEEE Transactions on Very Large Scale Integration (VLSI) Systems, Vol. 14, No. 10, pp. 1063-1074, October 2006.
27. A. Sheibanyrad, and A. Greiner, *Two efficient synchronous ⇔ asynchronous converters well-suited for networks-on-chip in GALS architectures.* Integration the VLSI Journal, Vol. 41, No. 1, pp. 17-26, 2008.
28. W. Ning G. Fen, and W. Fei, *Design of a GALS Wrapper for Network on Chip.* 2009 IEEE WRI World Congress on Computer Science and Information Engineering, Vol. 3, pp. 592-595, March 2009.
29. P. Liljeberg, J. Plosila, and J. Isoaho, *Self-timed communication platform for implementing high-performance systems-on-chip.* Integration the VLSI Journal, Vol. 38, No. 1, pp. 43-67, 2004.
30. R. Bashirullah, W. Liu, and R.K. Cavin, *Current-mode signaling in deep submicrometer global interconnects.* IEEE Transactions on VLSI Systems, Vol. 11, No. 3, pp. 406-417.
31. R. Bashirullah, *Reduced delay sensitivity to process induced variability in current sensing interconnects.* Electronics Letters, Apr. 2006, Vol. 42, No. 9.
32. A. Katoch, E. Seevinick and H. Veendrick, *Fast signal propagation for point to point on-chip long interconnects using current sensing.* European Solid-State Circuits Conference, 2002.
33. A. Katoch, H. Veendrick and E. Seevinick, *High speed current-mode signaling circuits for on-chip interconnects.* IEEE International Symposium on Circuits and Systems, Vol. 4, pp. 4138-4141, May 2005.
34. A. P. Jose, G. Patounakis and K. L. Shepard, *Near speed-of-light on-chip interconnects using pulsed current-mode signaling.* IEEE Symposium on VLSI Circuits Digest of Technical Papers, pp. 108-111, June 2005.
35. M. K. Gowan, L.L. Biro and D.B. Jackson, *Power considerations in the design of the alpha 21 264 microprocessor.* Proc. Design Automation Conference, 1998, pp. 726-731.
36. E. Nigussie, J. Plosila and J. Isoaho, *Delay-Insensitive On-Chip Communication Link using Low-Swing Simultaneous Bidirectional Signaling.* IEEE Computer Society Annual Symposium on VLSI, pp. 217-222, Mar. 2006.

37. M. -H. Oh and D. -S. Har, *A Novel Mechanism for Delay-Insensitive Data Transfer Based on Current-Mode Multiple Valued Logic*. PATMOS 2004, pp. 691-700.
38. V. Venkatraman, and W. Burleson, *Robust Multi-Level Current-Mode On-Chip Interconnect Signaling in the Presence of Process Variations*. Proceedings of the 6th International Symposium on Quality of Electronic Design, pp. 522-527, 2005.
39. T. Kuboki, A. Tsuchiya, and H. Onodera, *A 10Gbps/channel On-Chip Signaling Circuit with an Impedance-Unmatched CML Driver in 90nm CMOS Technology*. IEEE Asia and South Pacific Design Automation Conference, pp. 120-121, 2007.
40. N. Tzartzanis, and W. W. Walker, *Differential current-mode sensing for efficient on-chip global signaling*. IEEE Journal of Solid-State Circuits, Vol. 40, No. 11, pp. 2141-2147, 2005.
41. L. Zhang, J. Wilson, R. Bashirullah, and P. Franzon, *Differential current-mode signaling for robust and power efficient on-chip global interconnects*. IEEE 14th Meeting on Electrical Performance of Electronic Packaging, pp. 315-318, 2005.
42. A. Maheshwari, and W. Burleson, *Differential current-sensing for on-chip interconnects*. IEEE Transactions on VLSI Systems, Vol. 12, No. 12, pp. 1321-1329, 2004.
43. V. K. Venkatraman, *Design and Integration of Current-Mode On-Chip Interconnect Signaling in Nanometer Technologies*. PhD Thesis, University of Massachusetts Amherst, 2007.
44. W. J. Dally, B. Towles, *Route Packets, not Wires: On-Chip Interconnection Networks*. Proc. 38th Design Automation Conference, June, 2001.
45. M. Kamon, M. J. Tsuk and J. K. White, *FASTHENRY: A mutipole-accelerated 3-D inductance extraction program*. IEEE Transactions on Microwave Theory and Techniques, Vol. 42, No. 9, pp. 1750-1758.
46. A. Djordjevic, M. Bazdar, T. Sarkar and R. Harrington, *Linpar for Windows: Matix parameters for multiconductor transmission lines*. Software and Users Manual, Version 2.0, Norwood, MA: Artech House Publishers, 1999.
47. K. Banerjee, S. J. Souri, P. Kapur, and K. C. Saraswat, *3-D ICs: A novel chip design for improving deep-submicrometer interconnect performance and systems-on-chip integration*. Proceedings of the IEEE, Vol. 89, No. 5, pp. 602-633, 2001.
48. J. E. Sergent and A. Krum, *Thermal Management Handbook for Electronic Assemblies.*. The McGraw Hill Companies, 1998.
49. N. Lu, M. Angyal, G. Matusiewicz, V. McGahay, and T. Standaert, *Characterization, modeling and extraction of cu wire resistance for 65 nm technology*. Proceedings of the IEEE 2007 Custom Integrated Circuits Conference, pp. 57-60, September 2007.
50. K. Hirose and H. Yasuura, *A bus delay reduction technique considering crosstalk*. Proceedings of the conference on Design, automation and test in Europe (DATE), pp. 441-445, 2000.
51. T. Sakurai and K. Tamaru, *Simple formulas for two- and three-dimensional capacitances*. IEEE Transactions on Electron Devices, Vol. 30, pp. 183-185, 1983.
52. T. Verhoeff, *Delay-insensitive codes-An overview*. Distributed Computing, 3(1):1-8, 1988.
53. R. Venkatesan, J.A. Davis and J.D. Meindl, *Compact distributed RLC interconnect Models-Part IV: unified models for time delay, crosstalk and repeater insertion*. IEEE Transactions on Electron Devices, Vol. 50, No. 4, April 2003.
54. A. Narasimhan, S. Divekar, P. Elakkumanan, and R. Sridhar, *A low-power current-mode clock distribution scheme for multi-GHz NoC-based SoCs*. IEEE 18th International conference on VLSI Design, pp. 130-133, 2005.
55. L. Benini and G. Micheli, *Networks on Chips: Technology and Tools*. Morgan Kaufmann Publishers, 2006.
56. M. Krstic, E. Grass, F. K. Gurkaynak, and P. Vivet, *Globally asynchronous, locally synchronous circuits: Overview and outlook*. IEEE Design and Test of Computers, Vol. 24, No. 5, pp. 430-441, 2007.
57. E. Beigne, et al, *An Asynchronous Power Aware and Adaptive NoC Based Circuit*. IEEE Journal of Solid-State Circuits, Vol. 44, No. 4, pp. 1167-1176, 2009.
58. A. Sheibanyrad, I. Miro Panades, A. Greiner, *Systematic Comparison between the Asynchronous and the Multi-Synchronous Implementations of a Network on Chip Architecture*. IEEE Design, Automation and Test in Europe Conference and Exhibition, 2007, pp. 1-6, 2007.

59. A. J. Martin and M. Nystrm, *Asynchronous Techniques for System-on-Chip Design*. Proceedings of the IEEE, Vol. 94, No. 6, pp. 1089-1120, 2006.
60. W. J. Bainbridge, and S. B. Furber, *Delay insensitive system-on-chip interconnect using 1-of-4 data encoding*. International Symposium on Asynchronous Circuits and Systems, pp. 118-126, Mar. 2001.
61. W. J. Bainbridge, and S. B. Furber, *Chain: a delay-insensitive chip area interconnect*. IEEE Micro, Vol. 22, No. 5, pp. 16-23, 2002.
62. A. M. Pappu, X. Zhang, A. V. Harrison and A. B. Apsel, *Process-Invariant Current Source Design: Methodology and Examples*. IEEE Journal of Solid-State Circuits, Vol. 42, No. 10, 2007.
63. International Technology Roadmap for Semiconductors 2007, *http://public.itrs.net*.
64. D. N. Truong, et al, *A 167-Processor Computational Platform in 65 nm CMOS*. IEEE Journal of Solid-State Circuits, Vol. 44, No. 4, pp. 1130-1144, 2009.
65. Y. Hoskote, S. Vangal, A. Singh, N. Borkar and S. Borkar, *A 5-GHz Mesh Interconnect for A TeraFLOPS Processor*. IEEE MICRO, Vol. 27, No. 5, pp. 51-61, 2007.
66. S. Bell, et al, *Tile64^{TM} Processor: A 64-core SoC with Mesh Interconnect*. IEEE Solid-State Circuits Conference 2008.
67. D. Lattard, et al, *A Reconfigurable Baseband Platform Based on An Asynchronous Network-on-Chip*. IEEE Journal of Solid-State Circuits, Vol. 43, No. 1, pp. 223-235, 2008.
68. E. Salminen, A. Kulmala, and T. Hamalainen, *On Network-on-Chip Comparison*. Euromicro Conference on Digital System Design, pp. 503-510, 2007.
69. S. R. Vangal, et al, *An 80-Tile Sub-100-W TeraFLOPS Processor in 65-nm CMOS*. IEEE Journal of Solid-State Circuits, Vol. 43, No. 1, pp. 29-41, 2008.
70. M. B. Taylor et al, *The Raw Microprocessor: A computational fabric for software circuits and general purpose programs*. IEEE Micro Vol. 22, No. 2 pp. 25-35, 2002.
71. W. J. Dally and B. Towles, *Principles and Practices of Interconnection Networks*. Morgan Kaufmann, San Francisco, CA, 2004.
72. J. Kim, J. Balfour, and W. J. Dally, *Flattened Butterfly Topology for On-Chip Networks*. 40th Annual IEEE/ACM International Symposium on Microarchitecture, pp. 172-182, 2007.
73. J. Balfour and W. J. Dally, *Design tradeoffs for tiled CMP on-chip networks*. 20th ACM International Conference on Supercomputing, pp. 187-198, 2006.
74. L. Bononi, and N. Concer, *Simulation and analysis of network on chip architectures: ring, spidergon and 2D mesh*. IEEE Design, Automation and Test in Europe 2006 (DATE'06), Vol. 2, 6 pp., 2006.
75. M. Coppola, *Keynote lecture: Spidergon STNoC: the communication infrastructure for multiprocessor architectures*. In International Forum on Application-Specific Multi-Processor SoC, 2008.
76. W. J. Dally, *Enabling Technology for On-Chip Interconnection Networks*. Invited talk, 1st IEEE/ACM International Symposium on Network-on-Chip, May 2007 2007.$nocsymposium.org/keynote1/dally_n ocs$07.*ppt*
77. W. J. Dally, *On-Chip Interconnection Networks Low-Power Interconnect*. Special Session, International Symposium on Low Power Electronics and Design 2007, August 2007 *http://www.islped.org/X2007/DallyISLPED07.pdf*
78. U. Y. Ogras, and R. Marculescu, *"It is a small world after all": NoC performance optimization via long-range link insertion*. IEEE Transactions on Very Large Scale Integration (VLSI) Systems, Vol. 14, No. 7, pp. 693-706, 2006.
79. L. Benini and G. Micheli, *Networks on Chips: a new SoC Paradigm*. IEEE Computer, Vol. 35, No. 1, pp. 70-78, Jan 2002.
80. M. Krstic, E. Grass, F. K. Gurkaynak, and P. Vivet, *Globally asynchronous, locally synchronous circuits: Overview and outlook*. IEEE Design and Test of Computers, vol. 24, No. 5, pp. 430-441, 2007.
81. E. Beigne, et al, *An Asynchronous Power Aware and Adaptive NoC Based Circuit*. IEEE Journal of Solid-State Circuits, Vol. 44, No. 4, pp. 1167-1176, 2009.

82. S. -J. Lee, K. Kim, H. Kim, N. Cho, and H. -J. Yoo, *Adaptive Network-on-Chip with Wave-Front Train Serialization Scheme*. IEEE 2005 Symposium on VLSI Circuits, Digest of Technical Papers, pp. 104-107, 2005.
83. R. Dobkin, Y. Perelman, T. Liran, R. Ginosar, and A. Kolodny, *High Rate Wave-pipelined Asynchronous On-chip Bit-serial Data Link*. 13th IEEE International Symposium on Asynchronous Circuits and Systems, pp. 3-14, 2007.
84. C. K. K. Yang, *Design of High-Speed Serial Link Transceiver*. PhD Thesis, Stanford University, 1998.
85. M. J. E. Lee, *An efficient I/O and Clock Recovery for TERABIT Integrated Circuits Design*. PhD Thesis, Stanford University, 2001.
86. T. Bjerregaard, *The MANGO Clockless Network-on-Chip: Concepts and Implementation*. PhD Thesis, Technical University of Denmark, 2005.
87. E. Beigné, et al, *An Asynchronous NOC Architecture Providing Low Latency Service and its Multi-Level Design Flow*. 11th IEEE International Symposium on Asynchronous Circuits and Systems, pp. 54-63, 2005.
88. D. Lattard, et al, *A Reconfigurable Baseband Platform Based on An Asynchronous Network-on-Chip*. IEEE Journal of Solid-State Circuits, Vol. 43, No. 1, pp. 223-235, 2008.
89. R. Dobkin, R. Ginosar and A. Kolodny, *QNoC Asynchronous Router*. Integration the VLSI Journal, Vol. 42, No. 2, pp. 103-115, 2009.
90. E. Beigne, F. Clermidy, P. Vivet, A. Clouard, and M. Renaudin, *An asynchronous NOC architecture providing low latency service and its multi-level design framework*. 11th IEEE International Symposium on Asynchronous Circuits and Systems, pp. 54-63, 2005.
91. W. J. Dally and J. W. Poulton, *Digital Systems Engineering*. Cambridge University Press, 1998.
92. S. Borkar, *Designing reliable systems from unreliable components the challenges of transistor variability and degradation*. IEEE Micro, Vol. 25, No. 6, pp. 10-16, 2005.
93. P. Wang, G. Pei, and E. C. -C. Kan, *Pulsed Wave Interconnect*. IEEE Transactions on VLSI, Vol. 12, No. 5, pp. 453-463, 2004.
94. M. Chen and Y. Cao, *Analysis of Pulse Signaling for Low-Power On-Chip Global Bus Design*. 7th IEEE International Symposium on Quality Electronic Design, 6pp, 2006.
95. M. Bazes, *Two Novel Fully Complementary Self-Biased CMOS Differential Amplifiers*. IEEE Journal of Solid-State Circuits, Vol. 26, No. 2, pp. 165-168, 1991.
96. Y. I. Ismail, E. G. Friedman, and J. L. Neves, *Figures of Merit to Characterize the Importance of On-Chip Inductance*. IEEE Transactions on VLSI, Vol. 7, No. 4, pp. 442-449, 1999.
97. *International Technology Roadmap for Semiconductors 2008*. 0.5em minus 0.4em *http://public.itrs.net*.
98. B. Wong, A. Mittal, Y. Cao, and G. W. Starr, *Nano-CMOS Circuit and Physical Design*. IEEE Wiley-IEEE Press, 2004.
99. J. M. Rabaey, A. Chandrakasan, and B. Nikolic, *Digital Integrated Circuits: A Design Perspective*. Prentice Hall Ltd, 2 edition, 2003.
100. Li-Rong Zheng, *Design,Analysis and Integration of Mixed-Signal Systems for Signal and Power Integrity*. PhD thesis, Royal Institute of Technology(KTH), Sweden, 2001.
101. A. Deutsch, G. V. Kopcsay, P. W. Coteus, C. W. Surovic, P. E. Dahlen, and D. L. Heckmann, *Frequency-dependent losses on high-performance interconnections*. IEEE Transactions on Electromagnetic Compatibility, Vol. 43, No. 4, pp. 446-465, 2001.
102. A. V. Mezhiba and E. G. Friedman, *Power Distribution Networks in High Speed Integrated Circuits*. Kluwer Academic Publisher, 2003.
103. E. Rosa, *The self and mutual inductance of linear conductors*. Bulletin of the National Bureau of Standards, Vol. 4, pp. 301-304, 1908.
104. A. Ruehli, *Inductance calculations in a complex integrated circuit environment*. IBM Journal of Research and Development, Vol. 16, No. 5, pp. 470-481, 1972.
105. C. K. Cheng et al, *Interconnect Analysis and Synthesis*. John Wiley New York, 2000.
106. B. Young, *Digital Signal Integrity: Modeling and Simulation with Interconnects and Packages*. Prentice Hall PTR Upper Saddle River, NJ, USA, 2000.

107. A. Deutsch et al., *When are transmission-line effects important for on-chip interconnections.* IEEE Transactions on Microwave Theory Tech., Vol. 45, No. 10, pp. 1836-1846, Oct 1997.
108. A. J. Joshi, G. G. Lopez, and A. Davis, *Design and optimization of on-chip interconnects using wave-pipelined multiplexed routing.* IEEE Transactions on VLSI Systems, Vol. 15, No. 9, pp. 990-1002, Sept. 2007.
109. R. Dobkin, A. Morgenshtein, A. Kolodony, and R. Ginosar, *Parallel versus serial on-chip communication.* 10th International Workshop on System Level Interconnect Prediction, pp. 43-50, 2008.
110. *Semiconductor Industry Association. The International Technology Roadmap for Semiconductors, 2005 Edition. International SEMATECH:Austin, TX, 2005.* 0.5em minus 0.4em
111. D. Pamunuwa, H. Tenhunen, *Repeater insertion to minimize delay in coupled interconnects.* VLSI Design, Jan. 2001, pp. 513-517.
112. W. J. Bainbridge, W. B. Toms, D. A. Edwards, S. B. Furber, *Delay-insensitive, point-to-point interconnect using m-of-n codes.* Ninth International Symposium on Asynchronous Circuits and Systems, pp. 132-140, May 2003.
113. B. R. Quinton, M. R. Greenstreet, S. J. E. Wilton, *Asynchronous IC interconnect network design and implementation using a standard ASIC flow.* IEEE International Conference on Computer Design: VLSI in Computers and Processors, pp. 267-274, Oct. 2005.
114. B. R. Quinton, M. R. Greenstreet, S. J. E. Wilton, *Practical Asynchronous Interconnect Network Design.* IEEE Transactions on Very Large Scale Integration (VLSI) Systems, Vol. 16, No. 5, pp. 579-588, May. 2008.
115. *The International Technology Roadmap for Semiconductors(ITRS), 2007.*. Design, pp. 31, *http://public.itrs.net.*
116. O.S. Unsal, et al, *Impact of Parameter Variations on Circuits and Microarchitecture.* IEEE Micro, Vol. 26, No. 6, pp. 30-39, Nov.-Dec. 2006.
117. A. A. Mutlu, M. Rahman, *Statistical methods for the estimation of process variation effects on circuit operation.* IEEE Transactions on Electronics Packaging Manufacturing, Vol. 28, No. 4, pp. 364-375, Oct. 2005.
118. K. Bowman, J. Meindl, *Impact of within-die parameter fluctuations on the future maximum clock frequency distribution.* in Proceedings of Custom Integrated Circuits Conference, pp. 229-232, 2001.
119. H. -S. Wong, D. J. Frank, P. Solomon, C. Wann, and J. Wesler, *Nanoscale CMOS.* in Proceedings of the IEEE, Vol. 87, No. 4, pp. 537-570, 1999.
120. J. A. Croon, G. Storms, S. Winkelmeier, and I. Pollentier, *Line-edge roughness: characterization, modeling, and impact on device behavior.* in Proceedings of International Electron Devices Meeting, pp. 307-310, 2002.
121. P. Oldiges, L. Qimghuang, K. Petrillo, M. Leong, and M. Hargrove, *Modeling line edge roughness effects in sub 100 nanometer gatelength devices.* in Proceedings of the International Conference on Simulation of Semiconductor Processes and Devices, pp. 131-134, 2000.
122. S. Lakshminarayanan, P. J. Wright, J. Pallinti, *Electrical characterization of the copper CMP process and derivation of metal layout rules.* IEEE Transaction on Semiconductor Manufacturing, Vol. 16, No. 4, pp. 668-676, Nov. 2003.
123. X. Qi, A. Gyure, Y. Luo, S. C. Lo, M. Shahram, K. Singhal, *Measurement and characterization of pattern dependent process variations of interconnect resistance, capacitance and inductance in nanometer technologies.* 0.5em minus 0.4emin Proceedings of the 16th ACM Great Lakes symposium on VLSI, pp. 14-18, 2006.
124. D. Boning, *Pattern dependent characterization of copper interconnect.* Tutorial, International Conference on Microelectronic Test Structures, Mar. 2003.
125. L. Scheffer, S. Nassif, A. Strojwas, B. Koenemann, and N. NS, *Design for manufacturing at 65 nm and below.* IEEE 42nd Design Automation Conference, Tutorial 5, June 2005.
126. N. NS, *BEOL variability and impact on RC extraction.* IEEE 42nd Design Automation Conference, June 2005.
127. K. Bowman, S. Duvall, and J. Meindl, *Impact of die-to-die and within-die parameter fluctuations on the maximum clock frequency distribution for gigascale integration.* IEEE Journal of Solid-State Circuits, Vol. 37, No. 2, pp. 668-676, Feb. 2002.

128. N. N. Hoang, A. Kumar, P. Christie, *The Impact of Back-End-of-Line Process Variations on Critical Path Timing*. 2006 International Interconnect Technology Conference, pp. 193-195.
129. V. Mehrotra and D. Boning, *Technology scaling impact of variation on clock skew and interconnect delay*. International Interconnect Technology Conference, June 2001.
130. E. Demircan, *Effects of Interconnect Process Variations on Signal Integrity*. 2006 IEEE International System on Chip Conference, pp. 281-284, Sept. 2006.
131. J. Rabaey, *Low Power Design Essentials*. Springer Book Series on Integrated Circuits and Systems, 2009.
132. Y. Taur and T. H. Ning, *Fundamentals of Modern VLSI Devices*. Cambridge University Press, 1998.
133. N. James, P. Restle, J. Friedrich, B. Huott, and B. McCredie, *Comparison of Split-Versus Connected-Core Supplies in the POWER6^{TM} Microprocessor*. 2007 IEEE International Solid-State Circuits Conference, Power Management Papers, pp. 298-300.
134. H. Harizi, R. Haussler, M. Olbrich, and E. Barke, *Efficient Modeling Techniques for Dynamic Voltage Drop Analysis*. 44th ACM/IEEE Design Automation Conference, pp. 706-711, June 2007.
135. T. H. Morshed, et al, *BSIM4.6.4 MOSFET Model-Users Manual*. Department of Electrical Engineering and Computer Sciences, University of California, Berkeley, 2009.
136. Y. P. Tsividis, *Operation and Modeling of the MOS Transistor*. McGraw-Hill, New York, 1999.
137. N. Lu, M. Angyal, G. Matusiewicz, V. McGahay, and T. Standaert, *Characterization, Modeling and Extraction of Cu Wire Resistance for 65 nm Technology*. IEEE 2007 Custom Integrated Circuits Conference (CICC), pp. 57-60, Sept. 2007.
138. M. Ashouei, *Algorithms and Methodology for Post-Manufacture Adaptation to Process Variations and Induced Noise in Deeply Scaled CMOS Technologies*. PhD Thesis, Georgia Institute of Technology, Dec. 2007.
139. Arteris. [Online]. Available: http://WWW.arteris.com/
140. M. Coppola, C. Pistritto, R. Locatelli and A. Scandurra, *STNoC^{TM}: An Evolution Towards MPSoC Era*. NoC Workshop, DATE, Mar. 2006.
141. M. Coppola, R. Locatelli, G. Maruccia, L. Pieralisi, A. Scandurra, *Spidergon: a novel on-chip communication network*. International Symposium on System-on-Chip, Nov. 2004.
142. S. Xu, V. Venkatraman, and W. Burleson, *Energy-Aware Differential Current Sensing for Global On-Chip Interconnects*. 49th IEEE International Midwest Symposium on Circuits and Systems, MWSCAS'06, Vol.1, pp. 718-722, Aug. 2006.
143. J. Xu, and W. Wolf, *A wave-pipelined on-chip interconnect structure for networks-on-chips*. 11th Symposium on High Performance Interconnects, pp. 10-14, Aug. 2003
144. D. Schinkel, E. Mensink, E. A. M. Klumperink, E. van Tuijl, and B. Nauta, *A 3-Gb/s/ch transceiver for 10-mm uninterrupted RC-limited global on-chip interconnects*. IEEE Journal of Solid-State Circuits, Vol. 41, pp. 297-306, Jan. 2006.
145. E. Mensink, D. Schinkel, E. Klumperink, E. van Tuijl, and B. Nauta, *A 0.28pJ/b 2Gb/s/ch Transceiver in 90nm CMOS for 10mm On-Chip Interconnects*. IEEE International Solid State Circuits Conference (ISSCC), Digest of Technical Papers, pp. 414-415, Feb. 2007.
146. E. Mensink, *High-Speed Global On-Chip Interconnects and Transceivers.*. PhD Thesis, University of Twente, June 2007.
147. Haikun Zhu; Rui Shi; Chung-Kuan Cheng; Hongyu Chen, *Approaching Speed-of-light Distortionless Communication for On-chip Interconnect*. Asia and South Pacific Design Automation Conference, 2007, ASP-DAC '07, pp. 684-689, Jan. 2007.
148. Joonsung Bae; Joo-Young Kim; Hoi-Jun Yoo, *A 0.6pJ/b 3Gb/s/ch transceiver in 0.18μm CMOS for 10mm on-chip interconnects*. IEEE International Symposium on Circuits and Systems, 2008 (ISCAS 2008), pp. 2861-2864, May. 2008.
149. Medardoni Simone, Marcello Lajolo, and Davide Bertozzi, *Variation tolerant NoC design by means of self-calibrating links*. Conference and Exhibition on Design, automation and test in Europe (DATE 2008), pp. 1402-1407, Mar. 2008.

150. D. Mangano, R. Locatelli, A. Scandurra, C. Pistritto, M. Coppola, L. Fanucci, F. Vitullo, D. Zandri, *Skew Insensitive Physical Links for Network on Chip*. 1st International Conference on Nano-Networks and Workshops, NanoNet '06, pp. 1-5, Sept. 2006.
151. C. D'Alessandro, D. Shang, A. Bystrov, A. Yakovlev, and O. Maevsky, *Multiple-rail phase-encoding for NoC*. 12th IEEE International Symposium on Asynchronous Circuits and Systems (ASYNC), pp. 107-116, 2006.
152. P. B. McGee, M.Y. Agyekum, M.A. Mohamed, S.M. Nowick, *A Level-Encoded Transition Signaling Protocol for High-Throughput Asynchronous Global Communication*. 14th IEEE International Symposium on Asynchronous Circuits and Systems (ASYNC), pp. 116-127, Apr. 2008.
153. Y. Shi, S.B. Furber, J. Garside and L. A. Plana, *Fault tolerant delay insensitive interchip communication*. 15th IEEE International Symposium on Asynchronous Circuits and Systems (ASYNC), pp. 77-84, Apr. 2009.
154. D. Pamunuwa, L.-R. Zheng, and H. Tenhunen, *Maximizing throughput over parallel wire structures in the deep submicrometer regime*. IEEE Transactions on Very Large Scale Integration (VLSI) Systems,Vol. 11, No. 2, pp. 224-243, Apr. 2003.
155. L. -Z. Zhang, Y. Hu, C. -P. Chen, *Wave-pipelined on-chip global interconnect*. Asia and South Pacific Design Automation Conference, 2005 (ASP-DAC 2005), Vol. 1, pp. 127-132, Jan. 2005.
156. P. Cocchini, *Concurrent flip-flop and repeater insertion for high performance integrated circuits*. IEEE/ACM International Conference on Computer Aided Design(ICCAD), pp. 268-273, 2002.
157. R. Lu, G. Zhong, C. -K. Koh, and K. -Y. Chao, *Flip-flop and repeater insertion for early interconnect planning*. Proceedings of Design, Automation and Test in Europe Conference and Exhibition, 2002, pp. 690-695, Mar. 2002.
158. C. Lin and H. Zhou, *Retiming for wire pipelining in system-on-chip*. IEEE International Conference on Computer Aided Design 2003 (CAD 2003), pp. 215-220, 2003.
159. L. Scheffer, *Methodologies and tools for pipelined on-chip interconnect*. Proceedings of the 2002 IEEE International Conference on Computer Design: VLSI in Computers and Processors (ICCD02), pp. 152-157, Sept. 2002.
160. A. Lines, *Asynchronous interconnect for synchronous SoC design*. IEEE Micro, Vol. 24, No. 1, pp. 32-41, Jan./Feb. 2004.
161. J. D. Owens, W. J. Dally, R. Ho, D. N. (Jay) Jayasimha, S. W. Keckler, and L. -S. Peh *Research Challenges for On-Chip Interconnection Networks*. IEEE Micro, Vol. 27, No. 5, pp. 96–108, Sep./Oct. 2007.
162. L. Zhang, J. Wilson, R. Bashirullah, L. Luo, J. Xu, and P. Franzon *A 32Gb/s On-chip Bus with Driver Pre-emphasis Signaling*. IEEE Custom Integrated Circuits Conference, pp. 265-268, Sept. 2006.
163. A. Joshi, B. Kim and V. Stojanovi, *Designing Energy-Efficient Low-Diameter On-chip Networks with Equalized Interconnects*. 17th IEEE Symposium on High Performance Interconnects, 2009.
164. R. Ho, T. Ono, R. D. Hopkins, A. Chow, J. Schauer, F. Y. Liu, and R. Drost, *High Speed and Low Energy Capacitively Driven On-Chip Wires*. IEEE Journal of Solid-State Circuits, Vol. 43, No. 1, pp. 52-60, Jan. 2008.
165. E. Mensink, D. Schinkel, E. Klumperink, E. van Tuijl, and B. Nauta, *A 0.28pJ/b 2Gb/s/ch Transceiver in 90nm CMOS for 10mm On-Chip interconnects*. in Proc. IEEE International Solid-State Circuits Conference (ISSCC 2007), Digest of Technical Papers., pp. 414-612, 11-15 Feb. 2007.
166. B. Kim, and V. Stojanovic, *A 4Gb/s/ch 356fJ/b 10mm equalized on-chip interconnect with nonlinear charge-injecting transmit filter and transimpedance receiver in 90nm CMOS*. in Proc. IEEE International Solid-State Circuits Conference-Digest of Technical Papers, pp. 66-67,67a, 8-12 Feb. 2009.

167. J. -S Seo, R. Ho, J. Lexau, M. Dayringer, D. Sylvester, and D. Blaauw, *High-bandwidth and low-energy on-chip signaling with adaptive pre-emphasis in 90nm CMOS*. in Proc. IEEE International Solid-State Circuits Conference Digest of Technical Papers, pp.182-183, Feb. 2010.
168. R. Dobkin, M. Moyal, A. Kolodny, and R. Ginosar, *Asynchronous Current Mode Serial Communication*. IEEE Transactions on Very Large Scale Integration (VLSI) Systems, Vol. 18, No. 7, pp.1107-1117, July 2010.

Index

E.E. Nigussie, *Variation Tolerant On-Chip Interconnects*, Analog Circuits and Signal Processing, DOI 10.1007/978-1-4614-0131-5,

MIX
Papier aus verantwortungsvollen Quellen
Paper from responsible sources
FSC® C105338

If you have any concerns about our products, you can contact us on
ProductSafety@springernature.com

In case Publisher is established outside the EU, the EU authorized representative is:
Springer Nature Customer Service Center GmbH
Europaplatz 3, 69115 Heidelberg, Germany

Printed by Libri Plureos GmbH
in Hamburg, Germany